Regenerating Bodies

This exciting book examines how human tissues and cells are being exchanged, commodified and commercialized by new health technologies. Through a discussion of emergent global 'tissue economies', the author explores the social dynamics of innovation in the fields of tissue engineering and stem cell science. The book explores how regenerative medicine configures and conceptualizes bodies and argues that the development of regenerative medicine is a feminist issue.

In *Regenerating Bodies*, Julie Kent critically examines the transformative potential of regenerative medicine and whether it represents a paradigm shift from more traditional forms of biomedicine. The book shows that users of these technologies are gendered, and women's bodies are enrolled in the production of them in particular ways. So what is the value of a feminist bioethics for thinking about the ethical issues at stake? Drawing on extensive qualitative field research, Kent examines the issues around donation, procurement, banking and engineering of human tissues, and presents an analysis of the regulatory and policy debates surrounding these practices within Europe and the United Kingdom.

The book considers the claims that regenerative medicine represents exciting possibilities for treating the diseases of ageing bodies, critically assessing what kinds of future are embodied in tissue- and cell-based therapies. It will be of interest to a wide range of scholars and students within the social sciences, in health-technology studies, bioethics, feminist studies, and gender and health studies.

Julie Kent is Professor of Sociology of Health Technology at the University of the West of England in Bristol. Her research and publications examine the connections between ethics and regulation, the gendering of bodies, human-tissue use in the biosciences and the emergence of 'regenerative medicine'.

Genetics and Society

Series Editors: Ruth Chadwick, Director of Cesagen, Cardiff University; John Dupré, Director of Egenis, Exeter University; David Wield, Director of Innogen, Edinburgh University; and Steve Yearley, Director of the Genomics Forum, Edinburgh University.

The books in this series, all based on original research, explore the social, economic and ethical consequences of the new genetic sciences. The series is based in the Cesagen, one of the centres forming the ESRC's Genomics Network (EGN), the largest UK investment in social-science research on the implications of these innovations. With a mix of research monographs, edited collections, textbooks and a major new handbook, the series is a valuable contribution to the social analysis of developing and emergent biotechnologies.

Series titles include:

The Human Genome
Chamundeeswari Kuppuswamy

Community Genetics and Genetic Alliances
Eugenics, carrier testing and networks of risk
Aviad E. Raz

Neurogenetic Diagnoses
The power of hope and the limits of today's medicine
Carole Browner and H. Mabel Preloran

Debating Human Genetics
Contemporary issues in public policy and ethics
Alexandra Plows

Genetically Modified Crops on Trial
Opening up alternative futures of Euro-agriculture
Les Levidow

Creating Conditions
The making and remaking of a syndrome
Katie Featherstone and Paul Atkinson

Genetic Testing
Accounts of autonomy, responsibility and blame
Michael Arribas-Allyon, Srikant Sarangi and Angus Clarke

Regulating Next Generation Agri-Food Bio-Technologies
Lessons from European, North American and Asian experiences
Michael Howlett and David Laycock

Regenerating Bodies
Tissue and cell therapies in the twenty-first century
Julie Kent

Forthcoming titles include:

**Scientific, Clinical and Commercial Development
of the Stem Cell**
From radiobiology to regenerative medicine
Alison Kraft

Barcoding Nature
*Claire Waterton, Rebecca Ellis
and Brian Wynne*

Gender and Genetics
Towards a sociological account of prenatal screening
Kate Reed

Regenerating Bodies

Tissue and cell therapies in
the twenty-first century

Julie Kent

Routledge
Taylor & Francis Group

LONDON AND NEW YORK

First published 2012
by Routledge
2 Park Square, Milton Park, Abingdon, Oxon OX14 4RN

Simultaneously published in the USA and Canada
by Routledge
711 Third Avenue, New York, NY10017

Routledge is an imprint of the Taylor & Francis Group, an informa business

First issued in paperback 2014

British Library Cataloguing in Publication Data
A catalogue record for this book is available from the British Library

Library of Congress Cataloging in Publication Data
Kent, Julie, 1957–
  Regenerating bodies : tissue and cell therapies in the twenty-first
  century / by Julie Kent.
  p. cm. – (Genetics and society)
  Includes bibliographical references and index.
  1. Regenerative medicine – Moral and ethical aspects. 2. Stem cells –
  Research – Moral and ethical aspects. 3. Medical technology –
  Forecasting. 4. Feminism and science. I. Title.
  QH499.K45 2012
  174.2'8–dc23
  2011044835

ISBN: 978–0–415–68881–9 hbk
ISBN: 978–1–138–02011–5 pbk
ISBN: 978–0–203–33256–6 ebk

Typeset in Times New Roman
by Florence Production Ltd, Stoodleigh, Devon

Contents

Preface

This book engages with a series of questions about the implications of new and emerging health technologies that use human tissues and cells. Written from a sociological perspective, I set out to examine the meaning and significance of recent developments in biosciences and biomedicine that are targeted towards 'regenerating bodies'. I draw on research carried out over the past ten years, which began with some simple questions about women's experiences of breast implant surgery. Breast implant surgery (augmentation) can take place for a number of reasons: reconstruction following breast disease and mastectomy; to improve appearance owing to congenital or developmental 'deformity'; or for cosmetic enhancement. Such surgery 'reshapes' and, in a highly visible way, reconfigures the body. Implant technologies are inextricably social inventions, embedded within a complex cultural story about the meaning of breasts, the way clinical and social needs are defined, and how technologies develop and are distributed. Breast implants are technologies (medical devices) that have been commercially produced and marketed worldwide, provoking controversy about their safety and debate about how to regulate the industry (Kent 2003). Like other implant technologies such as hips or synthetic heart valves, breast implants are 'old technologies' that, although highly successful in many respects, have limitations in terms of their longevity and performance. As biomechanical components, the limitations of such 'replacement parts' have begun to seem outdated in recent years. In 2001, my collaborator Alex Faulkner described to me the development of a new type of implant that combined synthetic materials with 'living human cells', and we began to research the emergence of tissue engineering (TE) as a field of innovation within the biosciences. TE can be seen as 'an upgraded version of these earlier biomedical technologies', but can also be seen as going beyond the mechanistic approach to repair and being 'more concerned with the genesis of form – organogenesis – than the transplantation of already given forms' (Cooper 2008:111). This book draws on the research I did with Alex, Ingrid Geesink and David Fitzpatrick, investigating the regulation and governance of TE in Europe and the United Kingdom. I will refer here to data collected as part of that collaborative project. The study took place during 2002–4: we conducted a Europe-wide questionnaire survey of regulatory agencies (twelve out of seventeen countries responded); carried out sixty-three interviews among strategically placed regulators, European Commission (EC)/European Union (EU) officials, expert policy advisors, scientists, industry regulatory affairs and other staff, trade associations, consumer organizations and clinicians; collected extensive documentary materials and made detailed field notes at diverse meetings of regulatory policy networks that we attended. The focus of the research was regulatory and governance issues and the links between innovation and regulation; the ways in which risks associated with tissue-engineered products were constructed by different stakeholders; and how such technologies were being commercialized and brought to the market.

Since 2005, my interest in tissue- and cell-based technologies has taken a new turn, as I began investigating, with Naomi Pfeffer, the use of aborted fetal tissue in stem cell (SC) science. This research investigated how aborted fetuses are collected in the abortion clinic; the concerns that might influence women's willingness to donate an aborted fetus to SC research; the ethical, legal and policy frameworks governing the collection and use of aborted fetuses in research and therapies; the extent to which fetal tissue is used in SC science; why SC scientists value fetal tissue as a biological resource and research tool in SC science; and how fetal SCs are being translated into standardized therapies. We conducted participation observation of more than thirty scientific meetings; developed an extensive documentary analysis; carried out twenty-nine interviews with SC scientists, clinicians, tissue bankers, sponsors of research and activists; and convened six focus groups with forty-one women. Since then, I have continued to research what I call 'the fetal tissue economy' and ongoing international policy debates surrounding the use of human tissues and cells for research and therapies. This book therefore draws on extensive fieldwork and attempts to reflect on and synthesize those findings. Extracts from the interviews and focus groups are presented as transcribed verbatim with minimal editing.

TE and SC science are two approaches to what has more recently become known as 'regenerative medicine', which is associated with new epistemological communities and national and transnational industrial, scientific, policy and intergovernmental organizations and is linked to innovative practices in the clinical setting. I describe and analyse these developments through the book. Another overarching theme in the book is how relationships between the biological and the social are theorized and relate to regenerative medicine. In what sense does regenerative medicine mobilize specific understandings of the body, and how it may be acted upon or changed? How might we describe the relationships, for example, between the bodies of tissue donors and recipients of tissue- and cell-based therapies? These kinds of question lead to consideration of ethical issues relating to the exploitation of bodily tissues, bodily transactions and body commodification. They also point to ways in which gender difference may be significant in assessment of the impacts (potential and already realized) of contemporary 'tissue economies'. The book is organized around these themes, situated within a wider, global bioeconomy.

Chapter 1 introduces the notion that tissues and cells are being exchanged, commodified and commercialized in new health technologies. It describes the different sources of human tissue used in health technologies and introduces the science of TE and SC science. Using the example of the international skin business to illustrate the emergence of new types of tissue economy, it contrasts established practices of tissue transplantation and grafting with more recent developments in 'engineering skin'. In so doing, it draws on fieldwork interviews with tissue engineers and manufacturers of skin products. Tissues, cells and body parts are increasingly regarded as new types of commodity, with multiple uses in contemporary tissue economies (Waldby and Mitchell 2006). As venture capital is seen to invest in new techniques for extracting value from genetic material, cell lines and other body parts, the processes of commodification and commercialization may be seen as intensifying.

Chapter 2 explores the social dynamics of innovation and draws on data from my study of fetal-tissue collection and use as an example of a particular 'innovation environment'. It explores the idea that regenerative medicine represents a paradigm shift. A paradigm shift implies a new way of seeing and knowing that redefines how we think about bodies, illness and disease. Innovation is seen neither straightforwardly as a progressive development of new technologies, nor as distinct from social and political contexts; rather, technologies are socially embedded (Webster 2007). Diverse political cultures provide a context for innovation affecting policy, adoption and diffusion of new technologies (Jasanoff 2005). The chapter explores the political context for developments in SC science, tracing innovation in cell culturing techniques through to the standardization of methods and protocols for SC production in the United Kingdom and internationally. The role of the UK Stem Cell Bank for the governance of SC science is evaluated, and the chapter examines how new scientific tools are produced. Views of cells derived from aborted fetal tissue are also presented to illustrate the interlinking of scientific, technical and ethical issues in this field.

Chapter 3 examines the relationship between regulation and innovation and how new regulatory instruments and regulatory institutions developed in Europe. It focuses on the EU as the supranational state that interacts with national regulatory activities and provides a framework and context for the emergence of new regulation in public health and health-care products. This process of 'Europeanization' has significance for understanding policy formation, the shaping of legislation and governance process. It can be characterized by tension between national traditions and values and attempts to construct Europe-wide regulation that harmonizes member states' laws while, under specified circumstances, permitting national variations. Regulation and governance are seen as multi-layered processes that merit investigation from both the bottom up and top down. I discuss how ethical controversy around the use of human tissues and commercial exploitation of tissue- and cell-based therapies shaped international policy-making by drawing on fieldwork conducted at the time the new European Tissue and Cells Directive and the new Regulation of Advanced Therapy Medicinal Products were being developed. We see a convergence of the tissue-banking and TE sectors and attempts to define the boundaries between 'traditional tissues' and innovative technologies that use human biological materials. At the centre of the dispute were concerns about a generalized degradation of what it means to be human, the patentability of human life forms and promoting trade of new tissue- and cell-based therapeutic products.

Although the developments in European regulation described above provide an important context for, and framing of, international regulatory policy relating to regenerative medicine and the new tissue- and cell-based therapies, national (political) culture and history can also be seen as significant. The United Kingdom has had a high profile in this international arena and, some have argued, sought to lead debate. Chapter 4 provides a closer analysis of the dynamics of policy-making at the national level and reflects on what might be the specific features of the UK approach to regulating these technologies. It also examines how policy developments in Europe are being played out and implemented in the United Kingdom. I assess the claims made about the standing of UK research and policy in order to better understand the political dynamics in the UK setting. I argue that the British case must be understood in its specific historical and social context.

A number of important questions about developments in regenerative medicine are explored in Chapter 5. First, what kinds of body are enrolled in regenerative medicine as sources of 'raw materials' or, in industry parlance, 'starting materials' – the tissues, cells or body parts that provide the biological materials for processing? Second, how are technical systems and practices for processing, manipulating, storing and production of therapeutic product/ therapy gendered? Third, how are *users* of the technologies configured/ gendered, and how can patterns of access, distribution and availability of treatments be understood? Fourth, what is the transformative potential of these technologies? I suggest that the therapeutic benefits of therapies are likely to be different for men and women, given differences in disease patterns and treatment regimes. Women are also more heavily implicated within the emerging bioeconomy as donors, as many of the technologies being developed rely on the procurement of reproductive tissues (Waldby 2008, Waldby and Cooper 2010). The development of regenerative medicine is therefore an important feminist issue.

Chapter 6 examines what views of bodies emerge in regenerative medicine, and what the implications are for connections between bodies, identities, self and others. Rather than seeing a body as a unitary, 'skin-bounded' entity, more recently, within social theory, bodies and selves can be seen as multiple, relatively unstable, historically contingent, contested and malleable. This provides new analytical tools to think about emerging health technologies, including reproductive technologies, genetic technologies and cell- and tissue-based technologies. I assess the extent to which contemporary life sciences destabilize modernist ideas about the subject, showing how bodies have become understood as *fabrications*, as made, not born (Shildrick 1997). The intermingling of bodies, the *chimeric* qualities of maternal and other bodies unravel the notions of bodily integrity or well-defined bodily boundaries. With reference to focus-group and interview data, I discuss how the collection and use of aborted fetal tissue in research and therapies highlight the ontological problems of relationships between a woman and her (aborted) fetus.

The final chapter, Chapter 7, draws together the threads of the book and examines the claims made for regenerative medicine to prolong life and address the problems of ageing bodies. A critical assessment of such claims and the normative assumptions underpinning the innovation and regulation of new tissue- and cell-based therapies is presented. It reflects on how regenerative medicine 'remakes' life and death (Franklin and Lock 2003). The links between reproductive medicine and regenerative medicine are examined in relation to 'embryonic economies' (Franklin 2006a), and this is contrasted with the fetal-tissue economy. Whereas embryos symbolize 'life' and have been at the centre of disputes about personhood and the patenting of life, fetal tissue symbolizes both life and death. Fetal 'death' in the case of abortion (or miscarriage) releases the potential for life when tissue is donated for research or therapies. Whereas transplantation medicine redefined how the 'death' of a person was understood, regenerative medicine reframes ideas about the vitality of living tissues and cells. Commercial exploitation of tissue- and cell-based therapies frequently emphasizes their capacity to overcome the degenerative effects of ageing, even the promise of 'immortality', or at least extending life (Hall 2003). In conclusion, the chapter considers whether the political claims made for regenerative medicine are likely to have transformative effects and asks what kinds of future are embodied in tissue- and cell-based therapies and to what extent are the promises and expectations surrounding them likely to be realized?

Writing this book has been a process of reflection and synthesis of ideas, but I recognize that the story presented here is far from complete. While I modestly attempt to map out and discuss some aspects of the emergence of tissue and cell therapies in this century, we cannot yet know whether regenerative medicine will deliver significant social benefits.

Acknowledgements

My thanks go to all those who have supported and contributed to the genesis of this work. This includes those I have collaborated with over the past ten years or so – Alex Faulkner, Naomi Pfeffer, Richard Ashcroft, Ingrid Geesink, David FitzPatrick, Emma Williamson and Trudy Goodenough. Others who were part of the ESRC Innovative Health Technologies Programme and Stem Cell Initiative and whose work assisted the development of my own include Brian Salter, Catherine Waldby, Melinda Cooper, Sarah Franklin, Sarah Parry, Helen Busby, Christine Hauskeller, Paul Martin, Nik Brown, Neil Stephens, Peter Glasner, Clare Williams, Steven Wainwright, Andrew Webster and Margaret Sleeboom-Faulkner. Feminist scholars and friends who have also inspired and supported me include Susan Kelly, Maria Fannin, Helen Malson, Donna Dickenson, Michal Nahman and Lynn Morgan. I am grateful to the group in Public Health at the University of Copenhagen for the vibrancy of their intellectual work and for hosting my short study trip there – Mette Svendsen, Klaus Hoeyer, Sniff Nexoe and Lene Koch; to the ESRC for the funding of the two research projects on which this book draws, L218252058 and RES- 340–25–0002, and my fellowship, RES-350–27–004 during 2007–8, when much of this book was written. Thanks to those who agreed to be interviewed or to take part in focus groups as part of this research and to Geraldine Kinsella and Sophie Laws who both assisted in the 'Forgotten fetuses' project. My colleagues at UWE have maintained an interest in my research through tough and better times, especially Susan Hatt. Most importantly, I am forever indebted to Mike Kent for his unstinting love and patient understanding and for always managing to help keep me together. My biggest inspiration is the energy and enthusiasm for life of my mother Eve Hessey, to whom I dedicate this book.

1 Commodifying tissues and cells

The new tissue economies

Introduction

In Britain in 2006, news coverage of 'stolen body parts sold to the NHS' revealed that bone products used in dental and hip implantations represented a potential health risk to recipients, because the sourcing of bone had not been carried out in accordance with appropriate safety procedures. Following investigations by the US Food and Drug Administration (FDA) of a company supplying tissues and cells, it was revealed that the FDA had 'uncovered serious violations of the regulations governing donor screening and record keeping practices, as well as failures to follow their own standard operating procedures' to prevent contamination and reduce public-health risks.[1] Moreover the chief executive of the company and others were subsequently indicted for obtaining tissue from cadavers illegally and falsifying death certificates. The supply of human tissues was found to be illegal and unsafe, and companies using it in their products were implicated in a public-health scandal on both sides of the Atlantic. In the United Kingdom, the Medicines and Healthcare products Regulatory Agency (MHRA) notified hospitals of the ethical issues relating to lack of consent when the material was procured, but assured the public and clinicians that the bone filler product used in the United Kingdom did not represent a public-health risk, because it had undergone a sterilization process during preparation. It was possible to trace the material back to the source and to the recipients (though not all recipients had been informed of the source of the products used in their surgery). The supplier, Biomedical Tissue Services, was ordered by the FDA to cease trading.

In the same year, the UK Human Tissue Authority (HTA) reported, in its first regulatory alert, the use of research-grade materials being used in therapy – these were SCs derived from umbilical cord blood.[2] Research-grade SCs were distributed by Advanced Cell Therapeutics. Allegedly, these SCs, which were originally intended for research, had been distributed and used for human application, in other words for transplanting into patients. A British patient was treated in a Dutch clinic and subsequently suffered an adverse reaction. The company became the focus of an international fraud investigation and, at the time of writing, had ceased treating patients commercially (Sheldon 2006).[3] Safety concerns were raised again, this time because the cells had not been tested for use in patients. The company's wide-ranging claims about the efficacy of treatments using umbilical cord blood cells were regarded with scepticism and doubt by others in the field and were unsupported by appropriate clinical data.

These headlines, which provoke the kinds of emotional response one might expect, of horror, disbelief, disgust and concern, raise some interesting and important issues which are the focus of this chapter. In the first example, bone material, obtained from cadavers, was collected (harvested or procured) without consent and sold to a company that processed and distributed it to others in the commercial sector for use in various products. These products, which were widely available, were then implanted into patients who were unknowingly the recipients of stolen materials that were, at least in the view of the FDA, potentially unsafe and a risk to public health. What becomes evident is that there is a complex set of relationships underlying the exchange and circulation of these materials and the transformation of bone taken from a dead body into an implantable health-care product. In the second example, cells had been collected from an umbilical cord, that is, the living tissue of a new mother. Cells were extracted from the cord blood and processed (although it isn't clear what the nature of this processing was) and, although intended only for use in the research laboratory, were sold as a therapy to patients who unknowingly volunteered to undergo unproven treatment by being implanted with materials whose characteristics and qualities were uncertain and uncontrolled. In both cases, international and national regulatory authorities acted to stop these companies trading and investigated the risks to public health and safety (Sheldon 2007).

So, how do we make sense of what was going on here, and in what ways can a sociological understanding of health technologies assist us in analysing these events? To begin, I want to explore these 'tissue economies' and how tissues and cells circulate in a global bioeconomy (Waldby and Mitchell 2006). As part of this discussion, I aim to reflect on theoretical debate around the extent to which the human body is, or can be, commodified, and the implications of such a process. This will help us to think about commercial activities and the market for human tissue- and cell-based products (Andrews and Nelkin 2001, Dickenson 2002, 2005, Hoeyer 2007). This discussion focuses on the production of new health technologies and, although drawing on these wider debates, does not look in detail at either whole-organ transplants or blood and blood products. Neither am I especially concerned here with the use of biological materials in creating large databases, 'biobanks', for genetic or epidemiological research. Instead, my interest is in the production of technologies that use, for example, bone, cartilage, skin, neural cells and SCs, and are intended for the treatment of a range of diseases and bodily damage. These include degenerative diseases such as arthritis of the hip and knee joints; Parkinson's disease (PD); trauma such as sports injury to the knee, burns, stroke; and chronic conditions such as diabetes and chronic wounds. The range of conditions potentially treated by tissue- and cell-based technologies is wide and diverse; they affect large populations and are, in many instances, expensive to treat. There are then potentially lucrative markets for biotechnology companies proffering new therapies. In the cases above, profit was a key motivator for these companies who flouted safety and regulatory controls and potentially undermined the reputation and standing of others in the field.

Tissues, cells and body parts are increasingly regarded as new types of commodity, and there is extensive evidence of a global trade in them. While the donation of organs for transplant surgery has become relatively common-place, so too has the donation of gametes for use in assisted reproduction. As Scheper-Hughes notes, such practices are endorsed by physicians, surgeons and bioethicists 'as serving larger "altruistic" ends' (2001:31). Yet, despite assumptions that altruism underpins both decisions to donate organs and body parts and medical practices that use them, she and other commentators highlight the wider social and economic pressures that shape global trade and trafficking in organs and body parts. Her focus is primarily on 'illegal' but sometimes unofficially sanctioned trafficking, especially in contexts where extreme poverty leads individuals to sell their organs in order to buy basic supplies. She contrasts this with wealthy individuals able to buy organs or other health-care treatments.

The commodification of body parts has, as Lock points out, a long history, and the dissection of bodies for medical research and the development of medical knowledge is well documented (Lock 2001). With the emergence of genetic technologies, 'gene prospecting' has been criticized, as social groups are exploited for their genetic material, and 'divergent value systems' have emerged (Lock 2001:69). As venture capital is seen to invest in new techniques for extracting value from genetic material, cell lines and other body parts, the process of commodification is seen as intensifying. Moreover, the effects may be seen as gendered in relation to the commodification of ova and embryos for SC technologies (Dickenson 2002, Waldby and Cooper 2008). Waldby and Mitchell, in their assessment of human-tissue economies, argue that gift and commodity-exchange systems coexist and that, 'the values attributed to human tissues are always contingent on the particular cultural, ontological, epistemological and historical significance that they have for different actors in the various economies: donors, recipients, family members, medical staff, venture capitalists and so on' (Waldby and Mitchell 2006:34).

Moreover, according to them, tissue economies are sites for contesting and ordering these values. Biotechnology is implicated in the ordering of values and the creation of new objects from human tissue and in blurring the distinction between persons and things. These 'human boundary objects', it has been suggested, are produced through a process of commodification that results from a capitalist property regime and is generally regarded with moral repugnance, because it undermines notions of personhood and humanness. Yet, in his critique of the commodification hypothesis, Hoeyer suggests that new notions of property have emerged, associated with debates about the patentability of human biological materials (Hoeyer 2007).

At the centre of the debate about the circulation of tissues and the production of new kinds of tissue- and cell-based therapy are questions about both the ethical acceptability of commercial exploitation of body parts and political concerns about the distribution of the benefits and risks associated with commodification. In addition, there are important questions about how these technologies are socially shaped and what their transformative capacity is in terms of reshaping relationships between our bodies, our selves and others. These themes run through this book as I describe and analyse a number of 'tissue economies' (skin, cartilage and fetal tissue), how they are regulated (Chapters 3 and 4) and the emergence of notions of the 'regenerative body' (Chapters 5–7). My focus is on tissue-engineered health-care products and SC technologies as part of a broader development referred to as 'regenerative medicine'.

Emerging bioeconomies

According to Nikolas Rose, the twenty-first century can be characterized by a new form of biopolitics, a new, vital politics that in turn relates to intensified processes of the capitalization of life, new forms of biovalue and biological citizenship (Rose and Novas 2004, Rose 2007). He argues that,

> on the one hand, our vitality has been opened up as never before for economic exploitation and the extraction of biovalue, in a new bio-economics that alters our very conception of ourselves in new ways. On the other hand, our somatic, corporeal neurochemical individuality has become opened up to choice, prudence, and responsibility, to experimentation, to contestation, and so to a politics of life itself.
>
> (Rose 2007:8)

This, he suggests, is associated with a transformation in biomedicine to 'technomedicine', a new division of labour among specialists and new forms of expertise, new institutional organization of medicine and health care and capitalization of the biosciences. He identifies five 'mutations' from the biopolitics of the past, which include a shift from the 'molar body' to the molecular body; increasing emphasis on 'optimization' and the blurring of boundaries between therapeutic and enhancement technologies; the emergence of a new somatic ethics and subjectivity centred around the body; a multiplicity of expertise and therapists who act on the body and self; and the emergence of transformed bioeconomies that exploit the commercial potential of biological materials in innovative ways, resting heavily on a politics of hope and expectation (see also Brown and Kraft 2006) and leading to a reforming of relations between laboratory and factory.

This is less a matter of manufacture and marketing of falsehoods than of the production and configuring of truths. The reshaping of human beings is thus occurring within a new political economy of life whose characteristics and consequences we have yet to map.

(Rose 2007:32)

Encompassing the concept of biovalue used by Waldby (2002a) to characterize how bodies and tissues derived from the dead (and living) are redeployed for vitality of living, Rose uses it more widely to refer to vitality itself as a source of value. And, as illustrated by the work of Franklin (2007a), which traces the history of agricultural and sheep farming to modern cloning technology, he shares her view that the capitalization of biology is not new as such but seeks to capture latent value in biological processes in ways that are 'disembedding'.

Today a kind of 'dis-embedding' has occurred: vitality has been decomposed into a series of distinct and discrete objects, that can be stabilized, frozen, banked, stored, accumulated, exchanged, traded across time, across space, across organs and species, across diverse contexts and enterprises, in the service of bioeconomic objectives.

(Rose 2007:38)

Recognizing why such developments cause ethical concern, but in a critical appraisal of the diasporic notions of a posthuman future, Rose is more optimistic about the possibilities of this new, vital politics. He is reluctant to see either the direction of change as one sided or over-determined, but instead as presenting new spaces for contestation, conflict and political engagement.[4]

What is especially useful and interesting is that Rose's analysis situates emergent tissue and cell technologies within a much wider set of processes and practices that include genomic science, pharmacogenomics and psychiatry. My aim is much less ambitious, but complementary, in so far as I want to map out in more detail just what kinds of tissue- and cell-based technology are emerging; how specific tissues and cells are being sourced (harvested) and processed, how value is being created, and the circuits of exchange in which they are located; the possibilities they represent; and the conflicts that emerge in relation to them. To begin this task, I first want to describe how tissues are sourced for health technologies, before describing the development of the 'skin business' and new generations of viable skin products that are seen as at the forefront of regenerative medicine.

Sourcing tissues for health technologies

The social world of tissue banking is relatively unexplored. The history of tissue collection in the United Kingdom and the United States has been the focus of a recent study that illustrates the importance of tissue banks in processing and distributing cadaveric tissues such as skin, corneas and pituitary glands (Pfeffer 2009a). Other tissues, such as heart valves, bone and tendons, are also collected, stored and distributed. A central function of the tissue bank is to co-ordinate collection and to develop standardized procedures for ensuring the quality and utility of the tissues. The process whereby the tissues become standardized scientific objects or are produced as 'standardized therapeutic tools' is a complex social and technical process (Hogle 1995, 1996). Tissue banks are centrally concerned with the collection and distribution of human tissues and are important institutional structures within tissue economies. Their structure and function, however, are changing, and, although previously set up to distribute 'traditional tissues' from cadavers, their role has widened to include the processing and development of a wide range of 'products', including newer tissue- and cell-based technologies requiring greater degrees of manipulation or 'engineering'.[5] Moreover, the very concept of a 'tissue bank' has been redefined within the European Union to include both not-for-profit and for-profit 'tissue establishments', as we shall see later, in Chapter 3. Such transformations reflect wider changes in these 'tissue economies' and the exchange relationships that characterize them. The social worlds of tissue banking and the emergence of a commercialized TE industry can be seen as converging but also in tension. These tissue economies are shaped by, on the one hand, the history of tissue banking and, on the other, by the forging of new market relations and new forms of 'biocapital' (Rajan 2006).

Sourcing human tissue for therapeutic use is related to the development of a range of new knowledge and technologies that made blood transfusion and organ transplantation possible. *Blood* collection, which was important in times of war and national disaster, was linked to new techniques for collection, testing and storage. As these techniques developed, new types of blood product were derived, and the transfusion of whole blood became less common, undermining notions of the gift economy described by Titmuss (Busby 2006, 2010b, Waldby and Mitchell 2006). Dead bodies, or what Lock calls 'living cadavers' (Lock 2004),[6] have, since the 1960s, also been important sources of tissues used in transplantation medicine, which was reliant on the development of new surgical techniques and immunosuppressive drugs. Both *whole organs* (liver, heart, lung, kidney, spleen) and *tissues* (skin, bone, eyes) have been procured.

The 'procurement' or 'harvesting' of whole organs relies on a diagnosis of 'brain death', as well as technical and institutional arrangements for the co-ordination and distribution of organs. Death is legally, technically and officially defined, but its meaning is also contested and culturally specific. Brain death, though widely accepted, is revealed as problematic, in so far as it redefines the moment and shape of death and throws up fundamental questions about the relationship between the death of a person and biological death.

> The brain-dead patient-cadaver is a particularly complex hybrid, constituted from culture and nature while in transition from life to death, both person and nonperson, entirely dependent on a machine for existence.

> Technology – principally the artificial ventilator, but also other life-support equipment and procedures – is indispensable in the creation of living cadavers. Without them the machine-human hybrid could not exist.
>
> (Lock 2002b:40)

Technical practices, the actual process of organ removal, also vary, despite increased attempts to standardize them. So Hogle (1999) contrasts the 'craft-like' practices of surgeons explanting organs in Germany with the industrial approach in the United States, explaining this both in terms of different ideas about how to produce a quality organ and different priorities and also different cultural contexts. According to her, the US approach is different because it is more commercial. Transnational, national and local arrangements for organ procurement and distribution are also diverse, which influences exchanges and interactions between transplant teams, co-ordinators and recipients.[7]

Although organ retrieval usually takes place in the operating theatre, other tissues are likely to be collected in the mortuary. Organ retrieval from a brain-dead body, where the circulation is kept going using a ventilator and 'life-support' equipment, is carried out by a highly skilled surgeon. Afterwards, when the body is no longer being kept warm and pink, with tissues perfused with oxygenated blood (sometimes known as a heart-beating donor), it is transferred to the mortuary (when it becomes a non-heart-beating donor). Tissue procurement here will be done by trained, but not usually medically qualified, technicians and transported to the tissue bank for processing, storage and distribution. In some instances, however, the body itself will be transported to a special retrieval unit before being returned to the mortuary and prepared for burial or cremation. The tissues obtained may have a wide range of uses, for example in reconstructive surgery (or in treatment of burns), orthopaedic surgery (hip arthroplasty) or opthalmology (Pfeffer 2009a).

'Living donors' have also become more important in the effort to ensure a steady supply of organs for transplantation, especially kidneys (Department of Health 2007a). In Japan, most organ (kidney) transplantation has been from living donors, because the use of cadaver organs has been especially controversial (Lock 2002a, 2002b, 2004), but patients undergoing hip replacement are also important sources of bone products (Hoeyer 2009, 2010a).

A number of important themes emerge from the anthropological literature on organ transplants. First, the technologies of death are intimately tied into definitions of 'brain death' and are linked to procedures around seeking consent for organ donation. Second, diverse cultural meanings attached to the body and body parts influence social attitudes towards the desirability and acceptability of organ transplantation. Third, the alienability of body parts and body tissue is contested and underpins exchange relations between 'donors' and 'recipients'. Fourth, notions of subjectivity and bodily boundaries can be transformed by the experience of receiving an organ transplant. Fifth, body parts are seen as 'therapeutic tools' where they are used to treat diseased, ageing or degenerating bodies. What also becomes apparent from these studies of organ donation are that donors are 'selected' according to complex interactions of practical and technical criteria, but also social judgements about the 'worth' of certain types of body come into play.[8] Finally, studies of donor families and organ recipients suggest that organ transfer and the sharing of body parts can lead to new forms of kinship based on reinterpreting what 'blood ties' mean. Sharp asserts that the sharing of flesh and blood between donors and recipients can lead them to define themselves as related in particular ways, so that, rather than exchanging organs between 'strangers', the connections are personalized biological ties (Sharp 2007). In her analysis,

> we find that transplanted human organs are perceived by some parties as living things. They grant new life to recipients, who consider their personal surgical transformation to be a radical form of rebirth. In turn, a donor might well be perceived by surviving kin as having been reborn in the bodies of organ recipients, and spoken of openly and plainly as living on in several strangers. These paired rebirths generate unorthodox yet highly creative ways to confront grief as well as subvert the medical commodification of the human body.
>
> (Sharp 2007:74)

Relations between organ-donor families and recipients may be seen as creating new forms of sociality based around the shared or donated body part, a form of what has been called 'biosociality' – a term originally coined to describe new collectivities and identities within the genomic age (Bjorklund *et al.* 2003, Hopkins *et al.* 2005, Rose 2007). Such connections, although seen as positive in some stories, in others create feelings of obligation that cannot be fulfilled or satisfied, a debt that cannot be repaid, and the 'gift of life' becomes a tyranny because it cannot be reciprocated (Fox 1992).[9]

The donation and exchanges of *gametes* (sperm and eggs) by living donors is another well-researched area, which I refer to only briefly here, but will discuss in more detail later in the book, as it is especially important for analysis of the emergence of SC science and regenerative medicine. The ethical and social significance of *in vitro* fertilization techniques for the creation of human embryos since the 1970s has been extensively debated. Arising from animal-husbandry techniques, the move from the farm to the IVF clinic presented scientific, clinical and ethical challenges that shaped the politics of reproduction in modern societies. Access to treatment was strictly controlled, both by regulatory controls but also the ability to pay for expensive cycles of treatment (Pfeffer 1993). The conditions of procurement, especially of sperm, have in the past been relatively uncontrolled, whereas trading in ova, a much more invasive and riskier procedure, has (at least in some countries) been more strictly regulated. Payment of sperm and egg donors in cash or kind has been especially controversial in the United Kingdom and elsewhere, because it has been seen to undermine the values of altruism, promoting instead the commodification of human tissues and cells, though, at the time of writing, a

renewed debate about whether donors of these tissues should receive payment
in order to increase the supply has been initiated (Human Fertilisation and
Embryology Authority 2011). Where donor selection has been linked to
specific attributes being sought in the embryo, charges of 'designer babies'
have been directed at a consumerist view of reproductive technologies.
Safety issues relating to donor selection have recently received more
international attention, as we shall see in Chapter 2. Michal Nahman's (2006)
study of IVF clinics in Israel links the 'egg trade' to producing national
identities. She shows how Israeli women were less likely to accept ova donated
by Palestinian women but preferred donations from Romanian women, because
the latter group was believed to have desirable characteristics. However,
transnational markets in ova for reproduction or use in SC research have been
seen as potentially exploiting poorer women encouraged to sell their eggs
(Widdows 2009).

The collection of eggs for *research* rather than therapy has been especially
controversial in the United Kingdom and Europe. Two sources of egg donation
have been debated: first, donation of eggs by women undergoing IVF treatment,
as part of an 'egg-sharing' arrangement whereby the costs of their treatment
is reduced if they donate eggs for research. This is controversial because it is
seen as constituting a payment in kind for the eggs (Murdoch 2007, Roberts
and Throsby 2008). A second source of eggs for research is women who are
not undergoing treatment but who may be willing to undergo ovarian
stimulation for egg collection. This is referred to as 'altruistic donation', and
some critics argue it is unethical because it puts women's health at an
unacceptable risk, as the procedure is of no clinical benefit to them (Human
Fertilisation and Embryology Authority 2006).[10] The global flows of eggs, cells
and embryos have been traced to the flow of sheep-breeding technologies
(Franklin 2007a). Material exchanges of sheep and wool were historically tied
to innovation in the biosciences, knowledge exchange and technologies of
biological control. More recently, the emergence of 'embryonic economies'
to transfer embryos from the IVF clinic to the SC laboratory have increased
demand for a continual supply of donors and new forms of standardization
(Franklin 2006a).

Other tissues sourced at 'the beginning of life' are umbilical-cord blood and
aborted fetal tissue (though an aborted fetus can be seen as the end of 'life' –
more on this later). Cord blood has been valued as a source of haematopoietic
SCs in the treatment of blood disorders, and public cord banks set up to collect
and distribute it have been developed along the lines of other blood services.
However, the explosion of commercial cord-blood banks, designed to store
blood indefinitely for parents who deposit the cells as a kind of 'insurance'
against the future, is highly speculative and exploits promises of a future that
is as yet unknown. There are considerable uncertainties about whether the
stored cells will be of a sufficient quality (or quantity) to be of use, and it isn't
yet clear what kinds of illness or disease *might* be treatable using these cells.
Although bone-marrow transplantation and the use of haematopoietic SCs are

relatively well established, compared with other uses of SCs, these treatments follow protocols that are not necessarily readily transferable to the use of cells collected and stored in these commercial cord-blood banks (Brown and Kraft 2006, Brown *et al.* 2006b, Busby and Martin 2006a, Waldby 2006, Waldby and Mitchell 2006).

In the late 1980s, aborted fetal tissue was first used in experimental brain transplants in Europe, the United States, China and Mexico (Maynard-Moody 1995, Kent 2007, 2008a). Such experiments relied on a supply of aborted fetuses of an appropriate gestational age that could be co-ordinated with the preparation of a neurological patient to receive it. Then, logistical problems with obtaining a sufficient quantity of fetal neural material, of an acceptable quality, meant that, although some studies showed clinical benefits, this has not become an established practice. Ethical controversy also contributed to the cessation of trials in Parkinson's patients, although, perhaps unsurprisingly, in a small ethnographic study of these patients in Sweden, where the first trials took place, the patients themselves were grateful for the opportunity to receive fetal-tissue transplants (Lundin 2002). A new, multi-centred clinical trial using tissue from medically aborted fetuses in PD patients recently began.[11] Aborted fetal tissue has been used in SC science in the United Kingdom, and there is evidence of attempts to use it in commercial settings, as I shall discuss in more detail below (Kent 2007). Fetal skin and fetal bone are also thought to have potential benefits in the treatment of paediatric burns and fractures that don't reunite (Hohlfeld *et al.* 2005, Pioletti *et al.* 2007). In fetal medicine, fetal bone marrow and liver have been identified as sources of haematopoietic SCs, which are useful for *in utero* and post-natal transplantation treatment of congenital disorders (Golfier *et al.* 2000).

The use of non-human organs, tissues or cells – so-called xenotransplantation – is related to new forms of biotechnology and regenerative medicine and can also be seen as having a long history (Lundin and Idvall 2003). The use of pig body parts, for example porcine heart valves, and production of transgenic pigs to create transplantable organs have been a focus for research, despite complex technical issues relating to the safety of animal-tissue transfer to humans and ethical concerns (Sharp 2007). On closer investigation, the relationships and exchanges between human and non-human tissues and cells are extraordinarily complex. They include controversial experiments to transplant whole organs – baboon hearts, livers and kidneys – between the 1960s and 1990s in the United States. In 2000, a report on a pilot study of the transplantation of porcine cells into both patients with PD and others with Huntington's disease (HD) claimed the technique showed promise (Fink *et al.* 2000). Xenotransplants have been linked to fears and anxieties about whether the recipient might take on animal characteristics, in addition to the health and safety issues. Others have suggested that it is increasingly seen as an acceptable technology, at least among some groups (Lundin and Widner 2000, Lundin and Idvall 2003, Sharp 2007, Gaskell *et al.* 2010). However, definitions of xenotransplantation and categorization of xenotransplant products are not straightforward, as the mixing of animal and human cells in the Petri dish can also be seen as crossing species boundaries. Recent debate in the United Kingdom around the production of hybrids and chimeric embryos highlights technical variations of 'mixing' human and animal materials, and moral and regulatory concerns. Four types of hybrid, or chimera, were identified in these debates: human and animal gametes mixed to create embryos (true hybrids); embryos created by inserting a human nucleus into an enucleated animal egg (cytoplasmic hybrids); human embryos modified by the addition of animal DNA (transgenic embryos); or a human embryo altered by introduction of animal cells (chimera embryos) (Department of Health 2007b, Human Fertilisation and Embryology Authority 2007a, 2007b, Secretary of State for Health 2007). In addition, the use of mouse feeder layers and animal-derived serum for cell culturing is commonplace and has been the focus of scientific and regulatory discussions relating to the safety of tissue- and cell-based products (Shin and Griffin 2002).[12] These newer, 'hybrid' technologies can be linked to the emergence of hybrid regulatory institutions, as we shall see later (Brown *et al.* 2006a).

Engineering tissues

The new bioeconomies relate to the exploitation and capitalization of tissues and cells, and these newer technologies potentially create a demand for a wider range of tissue and cell types in order to create new health-care products. TE comprises a range of relatively innovative practices that have led to techniques of culturing cells and tissues to 'grow body parts' or 'regenerate bodies'. One definition is 'the persuasion of the body to heal itself, through the delivery to the appropriate site of cells, biomolecules and/or supporting structures' (Williams 1997,1999). Originating from pioneering work in the laboratory of Eugene Bell, Joseph Vacanti and Robert Langer at MIT in the United States (Lysaght and Crager 2009,[13] Langer and Vacanti 1995, Niklason 2001), early provocative images of 'earmouse' in the 1990s, which showed a mouse with what looked like a human ear growing on its back (but didn't contain human cells), suggested that three-dimensional tissue constructs (scaffolds) could be produced in the laboratory to grow replacement body parts. The effect of such images was to stimulate fears about the moral and social acceptability of xenotransplantation – the transfer of tissues and cells between species. However, the underlying principle was to 'engineer' or manipulate human tissue and cells *in vitro* and, in some cases, combine cells with synthetic materials or seed them on to three-dimensional structures for reimplantation into the body. Overcoming the technical requirements for generating complex human organs is still a long way off and, according to some commentators, an unlikely outcome, but simpler structures such as heart valves or blood vessels seem more possible, and, in 2009, a donor tracheal graft that was populated with the recipient's own cells was successful (Hollander *et al.* 2009).

There have also been efforts to regenerate tissue *in vivo* by, for example, creating a space within the outer layers of a long bone in which bone is encouraged to grow. The new bone can then be excised and reinserted into another site, enabling an *autologous* bone graft to be produced.[14]

> [This] approach is based on the manipulation of a deliberately created space within the body, such that it serves as an '*in vivo* bioreactor' wherein the engineering of neotissue is achieved by invocation of a healing response within the bioreactor space.
>
> (Stevens 2005)

So, there are different approaches to TE, which I will discuss in more detail in the next chapter, but, whereas the first is to expand cells and create tissues in the laboratory, the second approach intervenes in the body to create an 'artificial space', referred to as a bioreactor, in which new bone tissue is stimulated to grow. Interestingly, the body itself is conceived of here as a bioreactor, and it is precisely the ways in which these new technologies reconfigure bodies that are the focus of my analysis.

Stem cells

TE uses a variety of cells from the body (for example, keratinocytes, chondrocytes), but, since 1998, when human embryonic SCs were first isolated in the United States, SCs have been considered an especially rich biological resource. SC scientists aim to derive cell populations from a source and expand and proliferate the cells for implantation in a tissue-engineered product, as a cell therapy or for drug discovery (that is, where the cells mimic a disease or body function and can be used for testing drugs in the laboratory).

There are different sources of SCs – embryos, blood, bone marrow, umbilical cord, aborted fetuses and other (somatic) tissues of the body, including menstrual blood and placenta.[15] At present, it isn't known which source will prove most useful or valuable, but the human embryonic SC, which is pluripotent, is seen as the most plastic of cells because it is undifferentiated, or unspecialized, and can be used to produce any cell type of the body. The 'plasticity' of other cells and their potential to be differentiated into different cell types or to produce stable, replicable cell lines are under investigation and being debated within the scientific community. However, as we shall see, the scientific and political importance attributed to these cells is significant and has been seen both as heralding a new approach to body repair and regeneration and as potentially regenerating flagging economies (Cooper 2006a).

So, whereas in the next chapter I will explore the production of new scientific knowledge in these areas and the institutional arrangements that support it, here I want to begin to consider the shape of the market in engineered tissues and cell-based therapies (see also Bock *et al.* 2003, Martin *et al.* 2009). I will explore the features of the market for TE and SC technologies throughout the book, but begin here by focusing particularly on the skin business and the emergence of new types of 'tissue engineered skin'.

The skin business

Skin transplantation has a long history that has been traced back to early practices of nose reconstruction using a 'pedicle flap' in India and the development of 'free skin grafts' applied to burns and wounds during the nineteenth century in the West (Ang 2005).[16] There are three potential human sources of skin. First, the patient's own body, where skin is grafted from one part of the body to another by a plastic or reconstructive surgeon in the clinic (an autograft). Skin autografts are the 'gold standard' against which other 'skin substitutes' are evaluated (Kearney 2001, Shakespeare 2005). It became common practice for surgeons (particularly in burns units) to collect and store skin in their clinic fridges for use up to fourteen days later.

Second, skin can be sourced from a cadaver for use on another person (allograft). The procurement and processing of skin allografts raise different ethical and practical issues from those of autografts. Procuring skin for transplantation and grafting on to burns and other wounds has been underpinned by values that emphasize free and voluntary donation or gifting of tissue from cadavers or living donors. Like other forms of tissue donation, the collection of skin from dead bodies in the United Kingdom is a service performed in the hospital, as part of the retrieval process carried out by pathologists or technicians employed in a national network of pathology services and tissue banks. Screening of donors is required to reduce risk of cross-infection or disease transmission. Retrieval needs to be done within strictly specified time limits in order to ensure the skin cells are viable (Kearney 2005). Skin is harvested after circulation has ceased and it is collected under aseptic conditions using an electric dermatome and placed in a sterile container with a transport medium. Skin has three layers, and skin grafts can be of different thicknesses. Split-thickness skin grafts include the epidermal layer and different amounts of dermis; full-thickness grafts contain both epidermis and dermis (Hierner *et al.* 2005). It can then be stored either as non-viable 'biologic dressing', where cell death occurs but structural properties are retained, or as viable tissue (Pianigiani *et al.* 2005). The viability of allograft skin is around seven days, because deteroriation of the skin has begun prior to collection from the cadaver. Different methods of processing, preservation and storage are used in each case. Cryopreservation is one method of preservation for longer-term storage of viable skin (Kearney 2005), and glycerol will preserve non-living skin for up to five years.[17] Methods for the processing of human skin include techniques for washing, sterilizing and testing the material in order to avoid biological contamination and disease transmission and to promote safety and high standards of quality.[18]

Most European tissue banks were established in the 1970s and 1980s to facilitate more effective distribution of tissues.[19] Skin banks became larger regional or national skin or multi-tissue banks that 'operate according to Good Manufacturing Practice guidelines in rigorously controlled environments in line with defined protocols, hitherto an approach exclusive to the pharmaceutical sector' (Pianigiani *et al.* 2005:355).[20] They now interact internationally through networks that seek to promote uniformity in practices, and, in 2004, a harmonized regulatory framework was established by the European Commission (as I discuss in more detail in Chapter 2), which also impacted on the ability of surgeons to collect and store skin informally in hospitals, as this regulator explained to me in 2003, when new national guidance for tissue banks was introduced in the United Kingdom in anticipation of changes at the European level (Kent *et al.* 2006b):

> Well quite a lot of banks, quite a lot establishments didn't recognise they came under the Code of Practice [. . .] people like orthopaedic surgeons who feel they just want to store some bone, and plastic surgeons who treat patients who will take skin grafts and treat the patient, but they'll keep a bit of skin wrapped up in the fridge in case it fails so they can patch it afterwards – and all those people never dreamt for a moment that they would ever come under the Code of Practice because they didn't see themselves as tissue bankers. (R3, a UK regulator)

The development of tissue banks since World War II is linked to emergent technologies for producing standardized tissue 'products' (Cooper 2006b). Skin-tissue grafts have a wide range of applications including burns, which were a common clinical problem in wartime. Since the post-war period, burn patients have commonly been children or accident victims, although systematic epidemiological data on burns patients is incomplete (Brusselaers *et al.* 2010). There is also a large group of patients with chronic, difficult-to-heal venous and diabetic ulcers. In the Netherlands, the Dutch Burns Foundation set up a National Skin Bank in 1976, which became the Euro Skin Bank, but demand for skin outstrips supply in many countries (Freedlander *et al.* 1998).[21] Allograft skin acts primarily as a temporary biological dressing and stimulus for wound healing, but does not itself survive at the transplant site, owing to the immunological response from the recipient body (Snyder 2005). These are established practices that are being supplemented by newer TE technologies and cell therapies, which include acellular 'skin substitutes', autologous cell therapies and allogeneic (off-the-shelf) skin products for wound repair and regeneration. The production of these newer technologies is a commercial enterprise, reliant on private investment and exploitation of intellectual property. They are the focus of activities by both small, spin-out companies from university-based science departments (for example, Celltran) and global health-care companies (for example, Genzyme, Smith & Nephew), as I shall describe below. So, whereas the older, more established practices of skin collection, processing and storage in regional and national tissue banks were underpinned by service values and a distribution model for transplantation according to clinical need, the newer skin substitutes are associated with processes of commercialization and creating surplus value.

Engineering skin

Whereas the development of skin banks relates primarily to advances in burn management, the emergence of tissue-engineered skin products has been both for burns and the treatment of chronic, or hard to heal, wounds. At a scientific meeting in Bath in 2006, the representative of a well-known medical-device company suggested that there is a 'staircase to immortality' in the area of wound management and orthopaedics. He described the steps being taken towards shifting from an emphasis on skin replacement and repair to tissue regeneration. So, although the use of both autografts and allografts has a relatively long history (and it has been suggested that plastic surgeons are the earliest group of 'tissue engineers'), the scope for engineering skin, producing either synthetic skin substitutes (acellular) or products that incorporate viable skin cells, has led to the emergence of new types of innovative product (Supp and Boyce 2005).

There are two distinctive approaches to engineering skin – the first, *autologous* approach takes a patient's own cells and cultures and expands them in the laboratory, growing new skin for use on the same patient. Cultured autologous keratinocytes have been in use for many years, but, in recent developments, cells can be seeded on to a mesh, matrix or scaffold to provide structure and a method of delivering the cells to the site, and this can later be removed or be absorbed into the body, according to its design and properties (Harding *et al.* 2002, Hierner *et al.* 2005). Genzyme was one of the first to develop this technique, producing *Epicel*, a treatment for burns[22] that was found to assist with saving the lives of severely burnt and traumatized patients (Carsin *et al.* 2000). Genzyme is a multinational company with the ability to provide autologous services internationally. A biopsy of cells from the patient was sent by courier and plane to Genzyme's laboratories in the United States, and then sheets of skin were returned for use on the patient. In excess of 700 patients have been treated with *Epicel*.[23] Genzyme has another autologous cell-therapy product known as *Carticel*, an autologous chondrocyte implantation used in the treatment of cartilage defects (Kent *et al.* 2006a).

Another example of an autologous treatment, produced by a small spin-out company based in the north of England – Celltran, is *Myskin*. It arose from the collaborative work of Professors Sheila MacNeil and Robert Short at the University of Sheffield, who brought together the disciplines of wound healing and advanced polymeric materials to create new spheres of activity in tissue regeneration and repair. Celltran raised public and private investment and, although initially providing services at a local level, later distributed nationally to the NHS and expanded licence production across Europe and East Asia to reach worldwide markets (Press release, 2 October 2006). *Myskin* comprised a flexible, medical-grade backing polymer coated with a chemically controlled plasma polymer film that supports the growth of living keratinocyte skin cells. In order to obtain autologous cells, a small skin biopsy is taken from the patient (usually from the thigh) under local anaesthetic, transported to the laboratory and, like *Epicel*, it is co-cultured with irradiated murine feeder cells and bovine fetal calf serum. The cells are grown on a polymer sheet (PPS) and supplied in small circular discs in a sterile package for transport and delivery back to the patient.[24] The polymer film is engineered to promote cell growth and subsequent release when triggered by exposure to the wound. The time taken from biopsy to application of the cultured cells is one to two weeks. The discs are applied to the wound and covered with a dressing. Three or four days later, the disc is removed, and further applications of Myskin are applied if necessary. It was used in the treatment of burns and hard-to-heal wounds and diabetic foot ulcers. In response to concerns about the use of animal-derived feeder cells and bovine fetal serum, work was carried out to develop a serum- and animal-free method for cell culture (Bullock *et al.* 2006). Absence of reimbursement through national and insurance schemes was a continuing difficulty for expanding sales.[25]

In addition to the 'engineered skin', Celltran set up a small-scale tissue-bank service for local clinicians to store patient skin under conditions that complied with the new EU TCD and Human Tissue Act requirements (discussed below in Chapters 2 and 3). It was licensed by the UK Human Tissue Authority as a tissue bank and offered services to clinicians to bank both cadaver and autologous skin grafts. This represented an outsourcing of facilities that became too costly for local hospitals to provide, as the new regulatory requirements meant storing skin in a fridge at the clinic was no longer permitted, and, in order to meet the licensing requirements for a tissue establishment, costly new facilities would be needed in these hospitals. Celltran therefore represented a shift to a more highly regulated 'commercial' skin-bank and skin-product company that was initially based on a 'service ethic' but required both private investment and profitability for its continuation. In 2006, it merged with a Belgian company, Xcellentis, to enlarge its wound-care portfolio and extend its international business strategy, and, from 2008, a new company, Altrika Ltd, produced and marketed *Myskin*.

In 2007, Celltran (and subsequently Altrika Ltd) moved into the market for *allogeneic* skin products with *Cryoskin*, a cryopreserved sheet of human keratinocytes obtained from donor neonatal foreskins.[26] Allogeneic skin products such as *Cryoskin* and the first-generation products *Dermagraft* and *Apligraf*, also produced from neonatal foreskin, aimed to provide an 'off-the-shelf' cultured-skin product. In tracing the history of regenerative medicine, Paul Kemp, formerly employed at Organogenesis and founder of the UK company Intercytex, recounts a story of the emergence of TE and skin products from four Massachusetts laboratories of Green, Bell, Yannas and Vacanti (who collaborated with Naughton from Hunter College School of Health Sciences). Organogenesis was a company spawned from the Bell laboratory and led to the launch of Apligraf.[27] Dermagraft, invented by Gail and Brian Naughton, was developed by Advanced Tissue Sciences (ATS) and commercialized by Smith & Nephew. Apligraf was approved in the United States as a medical device and enjoyed initial success, but, when financial support was withdrawn by its partner company Novartis, Organogenesis filed for bankruptcy in 2002,[28] and, in the same year, ATS was bought by Smith & Nephew (Bouchie 2002).[29] Kemp said:

> As the new millennium dawned, the sector was in trouble. In the absence of commercially successful products or profitable outlicensing deals, tissue engineering companies struggled to make money and their overdependence on private investment became a real problem. The investor community tired of its new baby, turned its back on the sector and without substantial federally supported work, the bubble burst. [. . .] In hindsight the dramatic decline of the tissue engineering industry was inevitable, as it lacked a foundation of genuine scientific and commercial understanding . . . Tissue engineering needed more developmental biology and less engineering.
>
> (Kemp 2006b:658)[30]

In 2004, Organogenesis relaunched Apligraf on to the US market and, by 2007, had 230 employees and had treated 200,000 patients. Its focus was on automating and scaling up production, extending its distribution networks and developing a stronger business model, which included better links with the medical-device sector.[31] According to Geoff McKay, its CEO, who transferred from Novartis to lead the company, the future was bright: 'This moment in history will be remembered as a period when living, regenerative medicine products transform medicine. These products enhance, prolong, and in many cases, save lives' (CEO McKay, www.organogenesis.com, October 2007).

Dermagraft, meanwhile, was marketed by Smith & Nephew until 2006, when it was sold to the US-based company Advanced BioHealing. Smith & Nephew, it seemed, did not regard it as profitable.

Intercytex was set up in 1999 by Paul Kemp and commenced operations in 2000. It employed around seventy staff. Its head office was in Cambridge, UK, and it had a good manufacturing practice (GMP) clinical production facility, plus research and development laboratories in Manchester, UK, and additional laboratories in Boston, USA. Among other directors of the company was Mike Leek, who had formerly worked for Smith & Nephew running its tissue-repair programme and was involved in the joint venture between Smith & Nephew and ATS in the development of Dermagraft. Intercytex had four products in development; two were skin products. One was *ICX-PRO*, a wound-care product that comprised active allogeneic human dermal fibroblasts in a fibrin matrix. Like Apligraf and Dermagraft, the cells were derived from human foreskins in the United States. The product was manufactured in the UK GMP facilities and had a twenty-one-day shelf life. At the time of my research, it was in phase-3 trials in the United Kingdom, United States and Canada. It was designed for the treatment of venous leg ulcers and diabetic foot ulcers and was primarily being developed for the US market. A second skin product, *ICX-SKN*, was allogeneic skin replacement for wound closure (Boyd *et al.* 2007).

The company was initially set up with private venture-capital funding from the United Kingdom, Japan, Singapore and the United States and had licensing patents with universities in the United Kingdom, Japan and the United States. In January 2006, it was floated on the London Stock Exchange to raise additional capital and it subsequently received funding from the UK Department of Trade and Industry for the development of its plant to produce another product for the treatment of male-pattern hair loss. A business strategy was to fund research activities within university laboratories and then to bring the research in-house as the company developed. So, with a combination of outsourcing of laboratory work and in-house development, the company sought to avoid the financial difficulties encountered by Organogenesis and ATS and adopted a 'fully integrated business model' (Kemp 2006a).[32] However, following disappointing clinical-trial results for one of its products and other financial difficulties, the company was sold to a private holding company, but later relaunched in 2010 to focus on hair regeneration.

This brief introduction to the skin business points to emerging TE technologies and a market for new types of skin product. However, in 2003, a market survey suggested that there was little evidence of profitability for engineered skin and emphasized that commercial viability depended on accessing the wound-care market, providing clinical data that will assist in securing reimbursement, favourable comparisons with the costs of chronic care, and clear regulatory requirements.[33] As one academic scientist put it in an interview:

> I think tissue engineering products in the future won't be successful in the clinic if they only offer a small improvement over what's already out there. Classic example: Dermagraft may very well be a great product but it had to convince clinicians that it wasn't just a bit better than what they already had, but that it was ten times better to justify the cost. So the major limit is that what we do is going to have to be fantastically better than what's already out there on the market. Whereas in the pharmaceutical area, if a drug is just ten per cent better than what's already out there on the market, it has a good chance.
>
> (*UK Scientist*, 2003, S7)

Burns treatments do not have significant commercial potential in Europe (Husing *et al.* 2003).[34] Fetal skin cells – that is, skin taken from an aborted fetus – have also been used in clinical trials to treat children with burns, and further trials are planned (Hohlfeld *et al.* 2005). A patent has been filed by Hohlfeld and Applegate, but no commercialization of this technology has occurred yet.[35] Chronic wounds (including pressure ulcers, venous ulcers and diabetic ulcers) are costly and require intensive nursing care and community support (Harding *et al.* 2002).[36] Classical wound care uses traditional and advanced dressings; active management includes engineered skin. As a proportion of the total wound-management market, the use of engineered skin in 2002 remained relatively small (10 per cent; Bock *et al.* 2003). Skin defects can be treated using either classical methods, surgical procedures or split-skin transplantation – transplantation of engineered skin. Most are treated in the classical way, with low direct material costs, whereas engineered skin has high material costs (i.e. costs of production), which, perhaps surprisingly, it has been argued are similar for allogeneic or autologous sources.

So far, then, trading in skin products may be seen as reliant on more traditional skin grafts and wound-care management, despite the perceived potential of 'engineered skin'. Although these newer products illustrate the engineering principles applied to human tissue and have enjoyed limited commercial success, the evidence of a flourishing skin business is relatively weak. Autologous services that produce a customized product rely on high capital investment to build production facilities but do not produce high-volume sales for profitability. Allogeneic products also require high capital investment and have the potential to generate larger volumes of 'off-the-shelf products', but availability is limited by lack of reimbursement by health-care systems, which estimate the costs of these products as high compared with more established wound dressings. It has been argued that only where the high costs of nursing care for those with chronic wounds are included do the economic benefits of such products begin to come into view (Bock *et al.* 2003, Husing *et al.* 2003, Rowley and Martin, 2009).

In 2003, an estimated twenty-four products were on the market for skin replacement, and seven were in clinical trials. European companies favoured autologous products, whereas American ones favoured allogeneic products (Husing *et al.* 2003). No companies at that time had demonstrated profits from their skin products. A survey of the TE industry in 2004 suggested that there were a number of reasons for the underperformance and decline of the industry, and that the dynamics of change included a shift to smaller, SC companies and that 'skin, cartilage and other structural applications were especially hard hit' in contractions between the late 1990s and 2002 (Lysaght 2004:311). It identified lack of efficacy data, poor sales, high costs of production and regulatory approval and poor business skills as key problems, but argued that the industry remained viable and could be expected to expand and develop. Long-term failure was not inevitable, it said, but slow and uncertain beginnings were consistent with experience of other innovative medical technologies. A later study reiterated many of these earlier findings and identified similar barriers to the commercialization of regenerative medicine in the United Kingdom (Rowley and Martin 2009).

Conclusion

What does this discussion of the skin business and emerging engineered skin products tell us about the increased capitalization of bioscience? Certainly, we see that practices of skin-tissue exchange through the international skin-banking system are well established, and parallels can be drawn with other types of organ donation (skin is an organ).[37] In any case, cadaver skin used in the treatment of severe burns provides only a temporary dressing. Allogeneic skin products using donor cells continue to be present in chronic wounds, but not, apparently in burn sites; despite this, they have become accepted in the marketplace. Autografts have been the gold standard for skin repair, and cultured autologous cells (CEAs) have a good track record. Autologous, engineered skin products have showed promise in the clinic and some commercial success. The 'production' of living-skin equivalents and expansion of cells to create saleable and clinically efficacious treatments have come to typify the problems and possibilities for the TE industry. Creating biovalue and capitalization of this area of bioscience has created dynamic and changing relationships between tissue donors, cell biologists, engineers, pharmaceutical companies, venture capital, state sponsors, clinicians and patients in health-care systems. Yet, rather than an intensified, unidirectional process of capitalization, the evidence suggests a more complex, faltering process characterized by some enthusiastic scientists and entrepreneurial figures focused on 'getting the science right', but also on speculating on an uncertain futures market. In light of this, in the next chapter, I examine the extent to which regenerative medicine represents a paradigmatic shift in thinking about how to repair and heal bodies.

2 Regenerative medicine: a paradigm shift?

Regenerative medicine, propelled by the recent progress made in transplant medicine, SC biology, and related biomedical fields, is primed to expand the therapeutic armamentarium available in the clinical setting, and thereby, ameliorate disease outcome while reducing the burden of chronic therapy. This progress offers *a transformative paradigm* with curative objectives and goals to address disease management demands unmet by traditional (pharmaco)therapy. In particular, SC-based regenerative medicine is poised to drive the evolution of medical sciences from traditional palliation, which mitigates symptoms, to curative therapy aimed at treating the disease cause.

(Nelson *et al.* 2008:168; emphasis added)

The mission of the Regenerative Medicine Institute is to develop *a new and realisable paradigm for medicine*, utilising minimally invasive therapeutic approaches to promote organ and tissue repair and regeneration, rather than replacement.

(Regenerative Medicine Institute University of Galway,
Ireland, 2008)[1]

Introduction

There are a number of strategies for treating diseased, damaged and poorly functioning tissues or body parts. Interventions include drug therapies (for example, insulin injections for the treatment of diabetes; anti-coagulants for prevention of blood clots and heart disease; L-dopa to replace dopamine production in the brain), surgical excision or repair (such as mastectomy for breast cancer, removal of necrotic skin) and the use of implantable medical devices (replacement hip prosthesis, replacement heart valve) or transplanted tissues and organs (kidney, liver, skin). The production of medicines (synthetic substances) and devices made from a range of biomaterials is well established in clinical practice, as are the associated surgical procedures. The extracts above suggest that regenerative medicine has transformative potential, and institutions such as this one in Ireland aspired to develop 'a new and realizable paradigm for medicine', offering a different therapeutic approach to promote organ and tissue repair and regeneration. In this chapter, I want to explore the basis of these aspirations and claims and the dynamics of innovation that underlie them.

The dynamics of biomedical innovation have been a focus of attention for sociologists, who have highlighted the ways in which knowledge and technologies are co-constructed. They show how social networks, relations between the medical professions and the state, regulatory and governance regimes and links between academia and industry shape new health technologies. Accordingly, innovation is revealed to be intimately tied to sociopolitical processes in a way that requires analysis of sociotechnical relations, that is, the ways in which technologies are socially embedded and how social relations are materialized through technologies (Webster 2007, Faulkner 2009a). Such an approach also leads to an exploration of science and medicine as cultural practices that produce socially constructed meanings and objects – bodies, health and illness, therapies, health-care systems. A paradigm shift, in the Kuhnian sense, implies a new way of seeing and knowing that redefines how we think about bodies, illness and disease. Whether 'regenerative medicine', including TE and SC science, represents such a paradigm shift is open to debate. In so far as new knowledge (epistemology) is linked to new ways of being (ontology), such a shift also implies social transformations associated with technological changes. These are bold claims, which have promoted high expectations and contested futures (Brown *et al.* 2000, Brown and Webster 2004). Changing relations between the medical professions and the state have been linked to new knowledge and the rise of 'technoscience' (Pickstone 2000), and diverse political cultures provide a context for innovation affecting the policy, adoption and diffusion of new technologies (Jasanoff 2005). Moreover,

> The growing links between the state and the biotechnology, pharma-ceutical and medical device industries, and more recently corporations working in the information and communication technologies (ICT) sector have produced a more complex innovation environment, indeed not simply what (with particular reference to the United States) was originally called a 'medical–industrial complex' . . . but a medical–industrial–state complex.

(Webster 2007:43)

As we saw in Chapter 1, 'tissue engineering' can be traced to the work of Langer and Vacanti in Massachusetts, in the United States, in the 1980s. Massachusetts is itself an area that epitomizes changes in the social contract between universities and the state, and it was transformed into a site for new biotechnology companies that would facilitate technology transfer (Jasanoff 2005). The skin business is one area where techniques of expanding cells in the laboratory have been commercialized, though not as widely taken up in clinical practice as originally hoped. Indeed, evidence suggested that the TE industry has, in many respects, been a market failure, and that first- and second-generation products could not compete with more established clinical practices in health-care systems (Lysaght 2004, Kemp 2006a, 2006b, Martin *et al.* 2006, Mason 2007, 2007b). Increasingly, another vision of the future of biotechnology innovation has been to capture the potency of SCs, which are thought to offer the promise of 'regenerating' bodies and which can also be used in engineering tissues.

Following earlier support for SC research in the United Kingdom (Department of Health 2002), in 2005, the UK government expressed optimism about the future of SC research and regenerative medicine in response to the report from the UK Stem Cell Initiative (UKSCI). The UKSCI, chaired by Sir John Pattison, was established by Chancellor Gordon Brown with the aim of developing 'a ten year vision and costed strategy for UK stem cell research'. Pattison claimed that, 'innovation forms the backbone of the knowledge-based economy and SC research represents a substantial opportunity for future innovation in the life sciences.'

> The UKSCI vision is for the UK to consolidate its current position of strength in stem cell research and mature, over the next decade, into one of the global leaders in stem cell therapy and technology. The development of new stem cell therapies to treat conditions such as Parkinson's disease, diabetes and heart disease is one of the most exciting and captivating aspects of stem cell research. This is a vital and worthy aspiration for UK stem cell research and it remains important for the public and research community to be inspired, energized and driven by this long-term goal. Although it is reasonable to anticipate that some new stem cell therapies will be developed within the next decade, we must also accept that it is probable that this area will take several decades of small incremental advances in science and medicine to come to fruition. In this context, it is worth remembering that conventional pharmaceuticals take between 12–15 years of research and development to bring a product to market.
>
> (Pattison 2005:5)

The report sought to show both that there were established treatments using SCs – skin transplants (keratinocyte transplantation), bone-marrow transplantation and cornea transplants, dating from the 1940s, 1970s and 1990s, respectively – and new possibilities for exploitation:

> Most high-profile publicity for stem cell research currently focuses on the prospect of *regenerative medicine*. Here, the idea is that stem cells are grown to very large numbers in culture before *differentiation* into the desired therapeutic cell type. The cells would then be transplanted into patients in order to restore the function(s) lost due to accident or

disease. It is anticipated that regenerative medicine will one day be used to treat conditions such as Parkinson's disease, diabetes, coronary artery disease and spinal cord injury. In combination with other technologies, such as *tissue engineering*, it may even be possible to direct these cells to grow in the laboratory into highly organized tissues, or even organs for implantation into patients. Further embellishments include the use of *therapeutic cloning*, or *somatic cell nuclear replacement*, which may one day make it possible to generate cells which are genetic matches with the tissues of the patient, obviating concerns over immune system rejection of the stem cell transplant. If successful, stem cell therapy via therapeutic cloning would greatly contribute to personalized medicine.

However, it is important to recognize that stem cell research does not just encompass regenerative medicine. Furthermore, it is a misconception that stem cell research is new, or that we await the 'proof-of-principle' that stem cells will one day find therapeutic use. Indeed, the biological properties of stem cells have been exploited over the past several decades to develop a number of highly successful treatments.

(Pattison 2005:13)

Potential for exploitation includes the use of SCs for drug discovery and toxicology testing, and so, simultaneously, the promise of SCs was heralded as a new type of medicine – regenerative medicine – and innovation was linked with achievements of the past. In the UK context, this served a number of interests – it provided grounds on which to justify public and private investment in the future of SC research and, in emphasizing continuities with these histories and successes, sought to minimize the risks associated with these investments and promote expectations that new benefits to health could be delivered through 'public–private partnerships' in SC research.[2]

In November 2007, the scientific and popular press announced news that was heralded as another breakthrough in SC science. Two research teams, one in Japan and another in the United States, had successfully derived pluripotent SC lines from human somatic cells. The somatic cells, or 'adult cells', were fibroblasts from different sources – the skin on the face of a thirty-six-year-old woman, the synovium in the joints of a sixty-nine-year-old man, neonatal foreskin and an aborted fetus (Takahashi *et al.* 2007, Yu *et al.* 2007). Their achievement was to 'reprogramme' the cells by adding and removing genes, and the results were:

pluripotent stem cells that exhibit the essential characteristics of embryonic stem (ES) cells ... [moreover] such induced pluripotent stem cells should be useful in the production of new disease models and in drug development, as well as for applications in transplantation medicine once technical limitations (for example, mutation through viral integration) are eliminated.

(Yu *et al.* 2007:1917)

The news was welcomed by a UK-based company, ReNeuron, which saw it as endorsing its own approach to SC production (ReNeuron 2007). At that time, ReNeuron was awaiting the outcome of an Investigation of a New Drug (IND) application to the US FDA for approval to commence a clinical trial implanting cells from one of its cell lines into stroke patients to regenerate damaged brain tissue. The speed with which the company sought to identify itself with the achievements of the two scientific teams was striking and could be read as a shrewd marketing strategy to promote its own technology and a political move to encourage the US regulatory authority to approve its application to proceed to clinical trial. A more detailed consideration of these events is worthwhile, for what it reveals is a deeper connection between science in the Thomson laboratory in the United States and the commercial activities of companies such as ReNeuron in the United Kingdom or elsewhere. First, however, let us retrace the history of developments that preceded these events but that help to explain the significance of such pronouncements.

Continuities with the past: culturing cells

In her lucid account of the history of tissue culture, Hannah Landecker details the technical, practical and conceptual changes in biology in the early twentieth century. She argues that, 'the history of cell cultivation is the history of an approach to living matter that encompasses any specific example of biotechnical innovation such as stem cells' (Landecker 2007:7). Techniques of culturing cells, replicating them and sustaining them *in vitro* were central to new possibilities for experimental biology. *Plasticity*, she says,

> is an idea of living matter that is also a practical approach to it: substance may be radically altered without causing death. Plasticity is the ability of living things to go on living, synthesising proteins, moving, reproducing, and so on despite catastrophic interference in their constitution, environment, or form.
>
> (Landecker 2007:10)

The ability to grow cells outside the body was also seen, from the early twentieth century, as linked to biological time, a temporality that was not tied to the originating organism's life span but extended beyond it, and this gave rise to *immortality* as a technical term in biology. By providing us with her insights into the work of early-twentieth-century biologists, such as Jacques Loeb, Alexis Carrel and Ross Harrison through to Leonard Hayflick, Landecker shows how cells, as new types of scientific object, were created, together with the apparatus, culture media and imaging technologies that contained, stored, supported and monitored them. In the work of these earlier cell biologists, *reactivation*, *rejuvenation* and *regeneration* were claims made for their experiments in propagating and replicating cells and creating cell lines. As they honed their skills, they produced highly characterized (standardized) lines that could be distributed to other scientists. The famous HeLa lines, created from the tissues of a woman called Henrietta Lacks who died of cervical cancer, were 'immortalized' and were widely used in laboratories throughout the world (Skloot 2010). In the 1960s, Leonard Hayflick's work showed that, contrary to what had been believed since Carrel, somatic cells were not immortal but naturally died after a limited number of replications (known as the Hayflick limit), and that the immortality of the HeLa line should be understood as the effect of culturing cancer cells (see also Hall 2003). These effects helped to explain how cancer progressed.

As Landecker notes, 'good manners' required that researchers participate in informal arrangements for the exchange of cell lines, and state-funded institutions set up in the 1960s, such as the American Type Culture Collection (ATCC) in Maryland, facilitated exchanges on a larger scale. The establishment of such institutions, and the Tissue Culture Commission that predated them in 1946, facilitated the standardization of materials and practices. This in turn promoted extensive distribution networks among the biological community, mediated relationships between laboratories and ensured that standardized biological tools were used. Today, these exchanges continue, and ATCC supplied the fetal fibroblasts to the Thomson laboratory for their experiments in 2007.

These types of exchange are revitalized by new initiatives such as the International Stem Cell Initiative (ISCI), which aims to develop agreed markers for human embryonic stem cells (hESCs) and protocols for their production, including the use of standardized culture media (Andrews *et al.* 2005). At the 'hub' of this initiative was the UK Stem Cell Bank (UKSCB), which collected and prepared antibodies for distribution to the seventeen participating laboratories in eleven countries (the spokes) and received cell samples. Cooperation between laboratories was seen as necessary to achieve certain common goals in advancing SC science and facilitate certain epistemic outcomes – 'common tools and a common taxonomy'.

> Standards are seen as a collective good, or necessary evil, that will provide scientists with comparable data and thus a basis on which to make general claims about stem cells, instead of claims that only relate to one particular stem cell line.
>
> (Eriksson and Webster 2008:3)

In seeking to characterize and ensure the biological quality of hESCs, key questions need to be settled, such as what a stem cell is, what the effects of using different kinds of culture medium are, the effects of different 'passaging techniques' – that is, the transfer of cells from one bottle to another or plate to plate as they are expanded – and how you define 'pluripotency'. In answering such questions, standard setting is seen as a means of governing the field and managing uncertainties (Webster and Eriksson 2008).

Governing science through new public institutions: the UK Stem Cell Bank

The UKSCB was funded by the UK Medical Research Council (MRC) and the Biotechnology and Biological Sciences Research Council (BBSRC) and was established in 2003 at the National Institute for Biological Standards and Control. As Jasanoff (2005) points out, state sponsoring of science in the United Kingdom through the research councils is a part of the infrastructure that has forged relations between the state and science and has shaped science policy. According to one tissue banker I interviewed,

> The bank was set up largely because of the ethical issues surrounding the use of embryos for stem cell research. However, it was not set up just to curate embryonic stem cells. It's there to take in both embryonic, fetal and adult [cells]. (TB2, tissue banker)

It was, then, both a *political* response to international controversy surrounding the use of pre-implantation embryos for SC derivation and an attempt to address the scientific questions around standardization. Moreover, as the same interviewee explained, the UKSCB saw itself as 'a broker for the derivation centres and the commercial interests with the regulators and vice versa'. In order to maintain its 'independence', the bank was bound by the terms of its funding not to compete on a scientific or commercial level, nor to undertake fundamental SC research. Rather, a key function

> is to try and standardize methodologies but also to develop from that new ones that are applicable to a wide range of cell lines, and also to develop the characterization safety tests that we need to carry out in order to qualify and ensure that the line is now safe to use. (TB2)

The focus has been on embryonic SCs because there were very few somatic or fetal cell lines, and, although SCs could be isolated from a large number of tissues, 'they won't grow continuously or for a large number of passages. So they're not a cell line as such, they're almost like a primary cell . . . so there is a difference between a stem cell and a stem cell line' (TB2). Further explanation included a distinction between cells that age, go into senescence and die, and so are not immortal, and those that have a virus added to change the genome, enabling them to become immortal and 'behave more like a cancer cell than they do a normal cell'. In other words, an oncogene is used to switch on some genes and switch off others. These lines were considered unacceptable to the UKSCB (Medical Research Council 2005), but not simply because they were discounted as 'normal cells', but also because the management of the bank 'wanted to narrow down the range of cells taken into the bank', because they could only bank so many lines. Other restrictions on the types of line to be accepted in the bank included fetal neural SCs – because they're not clonal, nor immortal. However, what emerged was how ill defined terms and categories being used to establish the working practices of the UKSCB were, or, to put it another way, how scientific standards and biological categories were being co-constructed with the regulatory function of the bank as a social institution. Here, we see science in the making, new scientific objects being co-constructed and new boundaries being demarcated. The UKSCB can be seen as both working to stabilize these objects but also, at the same time, developing new trust relationships and social networks (Stephens *et al.* 2008).

Although standardization of methods and protocols was central to the work of the UKSCB and ISCI, ethical governance was also an explicit aim (see also Waldby and Salter 2008). Researchers working with embryos in the United Kingdom required a licence from the Human Fertilisation and Embryology Authority (HFEA) and were obliged to offer any lines derived from them to the bank. It was also a condition of the award of MRC funding. Approval of applications to deposit cells in the bank rested on the ability of researchers to demonstrate 'ethical sourcing', whether from the UK or abroad. Hence, in the wake of ethical controversy in Spain, cell lines from Valencia were not accepted by the UKSCB. In other words, in the United Kingdom, the bank worked to *legitimize* and validate hESC lines. Moreover this work of legitimization was closely tied to the social standing and legitimacy of the UKSCB itself. An ethnographic study of the UKSCB draws attention to how this was achieved by exploring how the bank's Steering Committee assessed the ethical and legal aspects of applications to the bank, and how the laboratory carried out the technical work that assessed the quality of the cell lines. Although, on the one hand, there were attempts to distinguish between these two dimensions of regulatory activity, on the other hand, the legitimacy of the UKSCB depended on the public seeing these two elements as inseparable. For,

> it is, however, only through the wider public perception of both of these elements brought together in the UK Stem Cell Bank as single unit that the Bank can secure simultaneously the legitimacy for its own future and that of hESC science more widely.
>
> (Stephens *et al.* 2008:46)

Through its work, the UKSCB and the ISCI sought to establish a network of laboratories and scientists whose lines are validated and legitimized and whose social position within this network is recognized. However, while the ISCI relied on the willingness of scientists to opt in and participate in the project, certain groups, because of their focus on 'non-embryonic' SCs, were in any case excluded. For example, fetal lines offered to the UKSCB were turned down both on ethical (political) and scientific grounds. So, the reach of the bank and the ISCI in ruling cells as eligible or ineligible for inclusion was significant, but those excluded fell outside their control and were relatively unregulated (Pfeffer and Kent 2007, Kent 2008a). Although, in some regulatory contexts, notably in the United Kingdom, distinctions between embryonic and adult SCs have frequently been taken for granted, these are problematic and highly contingent. Biological categories and concepts such as 'potency', 'plasticity' and 'immortality', in particular, have been contested (Eriksson and Webster 2008). (See Lakshmipathy and Verfaille (2005) for a discussion of plasticity, especially in relation to 'adult' SCs.)

Boundary making

In my research looking at the collection and use of fetal tissue in SC science, the dynamics of innovation and absence of standardization were evident in a number of ways. First, there was no agreement about what a SC is, and whether progenitor cells with the capacity to become different cell types exhibit 'stem-ness', as this scientist tried to explain:

> What the stem cell work has shown is that there remain, even in the adult brain, a very small number of progenitor cells that continue to undergo differentiation even in adult life. Whether you call them a stem cell depends on what your definition of a stem cell is. Most people would say strictly it's not a stem cell because it's a fairly late stage. It's still a progenitor. It's a cell with capacity to divide and with a capacity to become, probably the capacity to become several different cell types. So, how totipotent? How much capacity? You know, a conservative person or a purist would say a stem cell, you know, is only one of the cells of a blastocyst that can become any cell of the body and can be divided indefinitely. As you progressively go down the differentiation pathway, you get progressive restriction of lineages. (SS1, stem cell scientist)

Restrictions on what was acceptable to the UKSCB related to definitions of 'potency':

A: In the early days, there were a number of lines which the [UKSCB] steering committee didn't consider to be true stem cell lines. They were not considered – here we're going back to definitions again. And this is where that bit you had about the neural cell lines comes in. There's much debate over definitions but essentially a true stem cell is supposed to be pluripotent. Then some cells are considered to be totipotent and some are multipotent. And so the idea is that from the perspective of the embryonic stem cell, a true embryonic stem cell by definition should be capable of differentiating into cells of all three germ layers. So it can form neural cells, it can form cardiac cells, it can form cartilage cells. Now if you have something which is slightly further along the differentiated stage, it might be multipotent so it can form a number of different groups, but it can't form all of them. So the idea is that we [the bank] would stick to cell lines which were believed to be pluripotent.

Q: Doesn't that narrow it down to only embryonic stem cells?
A: Not necessarily, no. Because there is at least in theory that some of the human haematopoietic stem cells could be, if you go back far enough there should be cells that have the ability to move down a number of different lineages outside the blood lineage. (TB2)

If performance criteria were agreed as a way of measuring potency or plasticity, as the UKSCB and ISCI intended, then a strict definition of a SC might be produced, but instead the scientist above revealed the context of uncertainty about cell capacity or potency. In spite of these uncertainties, the public image of the embryonic SC was presented and promoted 'as a stable biological entity with a promising clinical future' (Eriksson and Webster 2008:57).

In contrast, such uncertainties have different consequences for how the SC derived from an aborted fetus is perceived. The derivation of these cells is more often hidden from public view, and they are highly unstable entities. Although scientists are frequently willing to engage with publics in explaining and elaborating aspects of their work, public discussion of SCs derived from aborted fetuses is limited, and information is difficult both to access and to comprehend for the non-scientist. Indeed, my research found that there was considerable confusion and obfuscation surrounding the different types of SC and their uses, perpetuated by the continuing production of policy statements and 'public information' leaflets that elided cell types and sources. For example, the MRC described embryonic SCs in the blastocyst and then said, 'the term adult stem cells can be misleading. In fact, so-called adult stem cells are found in babies, children and adults and even in the umbilical cord' (Medical Research Council 2006:6). Research Councils UK said there are two sources of SCs: embryos five days after fertilization and 'adult stem cells are found in many parts of the body such as the bone marrow, the eye, brain and muscle' (Research Councils UK 2007). Neither council statement referred to the aborted fetus. One scientist put it like this:

A: So there are broadly speaking two groups of stem cell biologists; of course, they're all very different disciplines, but two groups. There are those that are working on embryonic stem cells, and that's actually making stem cells from the embryo, and there are those that are working on adult stem cells, in other words, those people who are taking stem cells from adult tissues.
Q: Do you mean, by adult, me and you?
A: Yes, yes. Now, this is where the confusion comes in. I don't use the terms 'embryonic' and 'adult', and very rarely will you hear me talk about adult stem cells except to distinguish them from fetal stem cells from the same tissue, right? What I like to use is the phrase 'embryonic stem cells' and 'non-embryonic stem cells'. Because fetal stem cells, which is a

population, several populations of cells that people work on, are actually different to adult stem cells. (SS3)

Therefore, the term 'adult stem cell' says very little about the *source* of the cells, as they can be obtained from different types of body (baby, child and adult) and tissue (brain, liver, bone marrow and so on). 'Embryonic stem cells' is a term used to refer to the SCs found in the *pre-implantation* embryo at the blastocyst stage, even though 'embryonic' can also be used to describe the implanted, fertilized egg during its first eight weeks of development (Maienschein 2002). In the view of this scientist, 'fetal stem cells' are neither adult SCs or embryonic SCs, but something different. As a result, there is plenty of scope here for confusion and ambiguity, as he acknowledges. And, at the UKSCB:

We'll talk about being set up to curate adult, fetal and embryonic stem cell lines. So we make that distinction, that definition divided into those three. However, sometimes we'll talk about embryonic and somatic, and somatic including both adult and fetal. (TB2)

Evidently, the discursive production of fetal cells is highly unstable and contingent, and distinctions between cell types are highly problematic. Such classifications work in a number of ways (Bowker and Starr 2000). On the one hand, they serve to distinguish between cells from different sources and create new scientific objects, but, at other times, they obscure differences and render fetal cells (and abortion) invisible. They also have practical political effects, by setting limits on what might be accepted by the UKSCB or included in efforts to create standardized tools for SC research and therapeutic use. The UKSCB promoted a particular vision of the future where only certain types of SC – those that have been accredited and legitimated – were ethically acceptable and suitable for wider distribution and use in research or therapies, although the drawing of boundaries between embryonic SCs and other SCs that fall outside its Code of Practice were recognized in later iterations of it (Medical Research Council 2010).

Distributed networks, commercialization and therapeutic use of stem cells

The supply of biological materials and their distribution among the scientific community are associated with commercial production of pharmaceuticals, including vaccine and drug therapies. However, from a commercial perspective, it seems there was little to be gained from depositing SC lines in the UKSCB. In anticipation of the potential clinical and therapeutic use of embryonic SCs (and other lines), the UKSCB had, between 2003 and 2007, established itself as an accredited tissue bank (see next chapter), operating in accordance with GMP. Although lines received into the bank at the time were of 'research grade', which meant that they could not be used therapeutically, there was a hope that, at least in some instances, the lines could *become* clinical grade:

A: We didn't know and we still don't know whether it's possible to retrospectively switch an accredited line [. . .] to being therapeutic.

Q: So what are currently research grade might become clinical grade?

A: Yes. Now in the States they've done that with one line. The FDA have approved one line or are in the process of approving one line which was a research grade line to be used therapeutically. But the amount of work that has gone into doing that is probably more than it would be required to set up a CGMP facility. So it's possible in the future some of these lines might be used [therapeutically]. (TB2)

This would require more work, including a risk assessment of the safety issues arising from using mouse and animal feeder cells in maintaining the lines. However, although the UKSCB supplied cells as 'seed stock' to commercial laboratories, the scaling up of cells for therapeutic use, that is, production on an 'industrial' scale, was not an activity it would engage in. As a broker of SCs, it identified itself as part of a chain of custody and curator of master-cell banks from which it could create a supply for approved purposes (see also Glasner 2005). The operating costs of the bank were supported by public money, and no charges were implemented. One scientist explained the difficulties of funding the distribution process under normal circumstances:

It's a condition of getting a grant off the Medical Research Council or the Biotechnology and Biological Sciences Research Council that you make your material available but nobody tells you what happens the day you know, one hundred e-mails arrive asking for your cells because they're certainly not prepared to give you the resource to, for some technician to grow the damn cells up. And it's worse than that because you know you send the cells off, and then you get an e-mail back a week later saying they all died, we don't quite know . . . and they expect you to send another one and it can go on forever. (SS5)

So, although the UKSCB was seen as providing a useful service by distributing embryonic SC lines deposited there, its value to those using 'non-embryonic' SCs or those in the commercial sector appeared limited. Moreover, the commercial value of lines for therapeutic use depends on producing lines on an industrial scale, and, although commercial laboratories might value using standardized lines as research tools, it seems that there was little incentive for commercial laboratories to deposit lines that they themselves had created or to overcome the 'hurdles' required for them to be accepted. For example, ReNeuron set up licence agreements for distribution of its ReNcell 'human neural stem cell lines', via Millipore,[3] to other researchers for drug discovery (Donato *et al.* 2007). Therefore, there were parallel systems for distributing these biological materials – through the commercial sector and companies such as Millipore and, for selected hESC lines, through the UKSCB.

ReNeuron plc is an interesting case study of a small spin-out company focused on SC therapies. It is a UK-based, 'adult stem cell' therapy business, established in 1997 to commercialize technology discovered by scientists at the Institute of Psychiatry, King's College London (like Celltran, described in Chapter 1), and it maintains strong links with the institute. Although little information was publicly available about the sourcing of fetal tissue, the company sourced fetal tissue from Advanced Bioscience Resources in Alameda, California, an FDA-registered supplier of fetal tissue that also provides fetal bone marrow for haematopoietic stem cell (HSC) derivation (Mychaliska *et al.* 1998, Golfier *et al.* 2000), fetal eyes, thymus, pancreas and cord blood. The importation of this starting material was a business decision based on the need to meet the US safety standards for tissue and cells and 'donor-eligibility' criteria. It was thought that the risk of vCJD contamination from UK donors would be considered unacceptable, as the UK is regarded as a 'country of risk for vCJD'.[4] ReNeuron said their expansion technology obviated the need for further 'source material' (ReNeuron 2006). Until recently, it did not widely or explicitly advertise that its cells are fetal derived, but commonly referred to them as either 'adult' or somatic cells. A scientist I interviewed suggested,

A: I mean quite simply, they're trying to market the product as best they can. They're saying there's no distinction between fetal and adult, these are all . . . you know, these are not embryonic cells, these are adult cells. They don't want to draw a distinction, that's for marketing reasons, not to put too fine a point on it.

Q: Right.

Q: Is that because of the controversy associated with fetal? Are they trying to divorce it from its origin?

A: Yeah, they don't . . . they've got to market these cells in the US. They don't want to have to call them fetal cells. (SS5)

In its 2006 Annual Report, the company said it 'uses adult, or tissue-restricted, stem cells derived from non-embryonic human tissue sources' (p. 5) and that it used its *c-mycERTAM* technology to generate genetically stable neural stem cell lines. This technology platform has multi-national patent protection and is fully regulated by means of a chemically-induced safety switch. Cell growth can therefore be completely arrested prior to *in vivo* implantation.

In 2006, the company filed an IND with the FDA to commence initial clinical studies in the United States with its ReN001 SC therapy for chronic stroke disability. This represented the world's first such filing concerning a neural SC treatment for a major neurological disorder. The company hoped to begin trialling a stroke therapy based on animal studies and preclinical data (Pollock *et al.* 2006), with Professor Kondziolka, a neurosurgeon at the University of Pittsburgh Medical Center, as lead investigator. This phase-1 trial of ten patients, if approved, would last two years, but the company expected to extend its clinical trials from the United States into Europe.[5]

Interestingly, in updating its website in 2007, the company explicitly stated that it used 'somatic stem cells derived from the fetus' (August 2007), perhaps because it had become more risky not to identify sources. In addition to the treatment of stroke, the company had research programmes on the use of fetal-derived SCs for therapeutic treatment of PD, HD, Type 1 diabetes and retinal diseases. In April 2007, it announced additions to its patent estate, to include a European patent for 'the composition, manufacture and use of three key human neural stem cell lines' (Press release, 17 April 2007). It also had a cross-licence agreement with Stem Cells Inc. for the production of cell lines using its expansion technology.[6]

ReNeuron was regarded, in industry circles, as closer to the clinic with an SC therapy than many other companies, although there are now many other SC clinical trials ongoing across the world. However, the regulatory hurdles, described in more detail in the next chapter, and clinical concerns had to be overcome, and it wasn't until June 2010 that the company finally began its trial in the United Kingdom, following approval by the regulatory authorities there.[7] Moreover, from the results of the studies in the Yamanaka and Thomson laboratories, it was clear that those teams believed there was more work to be done before such induced pluripotent stem (iPS) cells could lead to a transplant therapy.

Innovation, transplantation medicine and stem cell science

In addition to the developments in cell biology, changes in clinical practice are also central to claims for an emerging 'regenerative medicine'. As already noted in Chapter 1, new surgical techniques, the management of immunological rejection of allogeneic tissues and histocompatibility tests for tissue matching were important innovations in the development of transplantation medicine from the 1950s. In addition, the history of blood services and bone-marrow transplants has been important and is documented elsewhere (Brown and Kraft 2006, Brown *et al.* 2006b, Kraft 2008, 2009). How bench science is translated into clinical practice has become a focus of attention, especially in relation to SC science, where improving both provision for, and efficiency in, translation is seen as vital for the development of SC therapies.[8] The Pattison report (Pattison 2005) identified the importance of 'translational research' and promoting and supporting clinical trials in the UK NHS, but, as Alison Kraft reiterates, 'Translation is not simply a case of transferring scientific knowledge from the research-base into the clinic. In practice, translation is neither linear nor unidirectional, but instead involves complex and evolving interaction between scientific, clinical, commercial and other groups.'[9]

She suggests that useful comparisons can be made between the ways in which HSCs have been used in cancer medicine and more recent 'HSC-based regenerative medicine'. She contrasts the work of translation in these two 'innovation environments', suggesting that the first can be characterized by continual and incremental change within an established therapeutic paradigm, whereas the second is highly novel, risk-laden and fraught with uncertainty (Kraft 2008, 2009).

A study of SC science as a future therapy for diabetes explored the interactions between bench and bedside in depth (Wainwright *et al.* 2006b). This study analysed the practical effects of scientists' accounts in constructing certain kinds of expectation about the trajectory from bench to bedside. Two main discourses on the prospects for translating SC research from the bench to bedside were identified: first, institutional influence on interactions between scientists and clinicians; and, second, SC science itself as a major barrier to potential therapies.

Institutional influences were identified as shaping interactions between scientists and clinicians in three ways. First, in the context of 'predictions of a revolution from conventional medicine to regenerative medicine' and government rhetoric supporting such a view, scientists felt obliged to demonstrate the clinical relevance of laboratory-based work in their grant applications. Second, they described a divide between the culture of medicine and biomedical science that hindered collaboration between scientists and clinicians and was attributed to differences in education, career structures and the organization of medical schools and science departments. Third, clinical scientists were perceived as especially pivotal in facilitating interdisciplinary research.

Since 2000, when the first islet transplants using cells from cadaver pancreas were performed using the 'Edmonton protocol', the possibility of treating diabetes with SC transplants has received greater attention. However, the scientists were seen as 'dampening down expectations' about the prospects of SC therapies for the treatment of diabetes. This was linked to problems of controlling embryonic SC growth and preventing the formation of tumours; problems associated with genetic modification to create insulin-producing beta cells and prevent immune rejection of implanted cells; and problems relating to the transferability of animal studies to human therapies. Accordingly, the authors concluded:

Our scientists see the target of ES-driven cell therapies as something that may be unachievable, except in very specific and limited areas. In contrast, they see the prospects for significant scientific breakthroughs from SCs in understanding basic cell and developmental biology as achievable. Perhaps there is a shift here, with some scientists now specifying expectations around SCs as scientific tools rather than medical therapies.

(Wainwright *et al.* 2006b:2061)

A clinical scientist I interviewed who has a specialist interest in diabetes was concerned about the relative risks and potential benefits of cell transplants. He explained that patients transplanted with cadaver islet cells had been those with diabetes that was particularly difficult to manage, and it was too early to conclude that cell transplants following the Edmonton protocol were a successful 'cure' for diabetes. Drawing parallels with the early bone-marrow transplants and contrasting the risks of treating cancer or heart failure with SCs and diabetes, he said:

A: So that, it's a risk–benefit analysis [. . .] Now, what's concerning to me, if I put my clinical hat on, stem cells [. . .] all get lumped into one basket. You've got a chronic disorder with diabetes, now insulin, everyone says you know, insulin's not a cure, it's a treatment. You're absolutely right, but it's a bloody good treatment and people don't die because they lack insulin in the immediate term after being diagnosed with Type 1 diabetes. Now [. . .] you may reduce your life expectancy by five to ten years. You still get a decent length of reasonable quality life if not good quality life, and therefore you have to be very careful of how you intervene with stem cells. That's a totally different kettle of fish from someone who is going to be dead within six months to a year because they've end-stage cardiac failure. (SS2)

He was also cautious about the time it might take to develop an SC therapy for diabetes and felt strongly that the tendency to lump all types of SC treatment together was *dangerous* because it raised expectations of patients, so some might think that there was no need to continue managing their diabetes using insulin. In his view, the cynical 'rebranding' of more established therapies, such as bone-marrow transplants or skin transplants, as 'stem cell therapies' by policymakers and some supporters produced unwanted effects – hype and unrealistic expectations about how long it could take to bring new therapies into the clinic. He said many years' work would be needed in basic science before a cell therapy for diabetes could emerge. He viewed a clinical trial in Brazil, using autologous HSCs in newly diagnosed patients with diabetes, some of whom were children, with scepticism and concern (Voltarelli *et al.* 2007).

Innovation in neuroscience using fetal tissue

In another 'innovation environment' – the neurology clinic – fetal tissue transplants have been important but are still highly experimental and have not become well established in the clinic. They have, however, been seen as important precedents for future SC transplants in the brain to treat neuro-degenerative diseases such as PD and HD. In this field, innovation may be characterized by continuing uncertainties but has been largely driven by clinicians working in close collaboration with neuroscientists. Cell transplantation techniques have used fetal tissue derived from aborted fetuses. Pioneering work to develop an understanding of how nerve fibres in the brain grow and connect, and experimental work developing cell transplantation in animal models (rat and primate) took place in the late 1970s and early 1980s. In the late 1980s, the first human transplants were carried out using a stereotactic surgical technique to inject human fetal cells into the patient's brain. The Swedish team, led by Olle Lindvall and Anders Bjorklund and based at Lund University, was at the forefront of this work, together with a Cambridge-based scientist, Steve Dunnett, who worked with them and went on to set up his own laboratory in Cambridge and later in Cardiff. He works closely with Roger Barker and Anne Rosser (a clinical scientist who works in the laboratory and neurology clinic). In Lund, permission to transplant fetal neural cells from aborted fetuses (8–10 weeks) into patients with PD was obtained, and the first transplants were carried out in 1987 (Bjorklund *et al.* 2003, Lindvall and Bjorklund 2004). In the United Kingdom, the following year, another researcher and neurosurgeon, Edward Hitchcock, also attempted neural transplants in twelve patients, using cells taken from second-trimester-aborted fetuses (11–19 weeks), in Birmingham. Other studies in the 1980s were carried out in China, Cuba and the United States (Hitchcock 1994). In the United States, the first trials were led by Curt Freed, whose team has performed over sixty fetal dopamine cell implant surgeries since 1988.[10] President Reagan's moratorium on federal funding of fetal-tissue research meant that Freed charged patients to take part in his early trials (Maynard-Moody 1995). By 2004, an estimated 350 patients with PD had received transplants (Lindvall and Bjorklund 2004).

Cells from fetal brain were found to have special qualities to grow and make connections, as this scientist explained:

> For brain, not withstanding modern views about stem cells, essentially, our nervous system is laid down in early development and we have it for life. There is not a turnover of nerve cells. If there was, you wouldn't be able to learn and remember anything because the connections would be replaced and lost. So it turns out that just from empirical studies, one of the breakthroughs that they achieved in getting cell transplantation in brain of adult mammals to work, is that the donor cells have to be from the fetal brain. They have to be the precise cell type that you are trying to replace

[. . .] the cells can survive, not only survive transplantation, but they then continue to grow and develop and integrate in the host brain. (SS1)

What emerges in this account is the notion of a 'replacement model', where replacement dopamine-producing cells are sought and repair is achieved by replacing damaged cells with transplanted ones. Different cell types are present in the developing fetus and can therefore be harvested at different stages of development. In order to obtain dopamine-producing cells, aborted fetuses of at least 6–8 weeks' gestation are needed.

The scientific literature indicates different views about the research protocol – the gestational age of the fetus from which cells are sourced, the dissection and preparation of the cells for transplantation, screening of donors and surgical techniques are debated. The notion of 'stem-ness', that is, what counts as an SC, is uncertain and contested, as we have seen already, and, for some, only the pluripotent cells derived from the blastocyst could be regarded as true SCs. The early fetal-tissue transplantations used 'primary cells' (sometimes referred to as 'embryonic dopamine neurons' (Freed *et al.* 2001), although there were also issues of quality control. Criticisms emerge from the literature about the extent to which some research teams were successful in isolating specific cell types for transplantation, that is, the 'purity' of the cells transplanted, and whether a 'cocktail' of different cell types were used. According to current standards, cell quality and viability need to be ensured, but couldn't always be adequately verified in some of this early work.

Other problems related to access to fetal tissue. The problems of supply included both the need to source tissue 'ethically', that is, with the formal consent of women 'donors', and practical problems of obtaining sufficient volume of tissue for a single transplant procedure (itself also a quality-control issue). The development of new ethical guidelines for the use of fetal tissue in transplants dates from this period (Polkinghorne 1989, Boer 1994).

The problem with the primary cells is availability, and it's not just *limited supply*. But if, as we are, at present, you're restricted, you're dependent for transplantation on cells derived from elective abortion, you can never establish the quality control that is, that you want either in terms of just good medical practice and increasingly in terms of regulatory conditions, and hence, the interest, the booming interest in stem cells is, whether you are interested in heart transplants but can't get heart cells or in neural cells and can't get fetal cells, the whole stem cell field has boomed. Partly because there is an interest in developmental biology but much more because suddenly, the incredible potential is the potential for *cells on demand* for an incredibly diverse range of human diseases, from heart attacks to diabetes to Parkinson's. (SS1; emphasis added)

In addressing the question of whether transplantation of primary cells from human fetal neural tissue could ever become a viable treatment for PD, these scientists concluded:

For us the answer is no. Whatever the outcome of small-scale trials and whatever improvements in techniques are made, transplantation of human fetal tissue can never be a large-scale treatment for all patients with advanced PD. The main reason for this rather pessimistic view is that the procurement of human fetal tissue requires a chain of expertise that cannot be 'industrialised'.

(Bjorklund *et al.* 2003)

The time, skill and logistics of procuring fetal tissue were obstacles to expanding the practice.

Problems evaluating these studies relate to the methods for assessing patients post-operatively and agreeing protocol for measuring the effects and outcomes of the transplantation (Freed *et al.* 2001, Gordon *et al.* 2004, Piccini *et al.* 2005). In 2003, in the United States, reported findings of side effects from fetal primary-cell transplants and some concerns about the use of sham surgery in double-blind trials led to the suspension of further transplants for PD and a shift in focus to the potential of SCs for treatment as a way of producing 'cells on demand' (Check 2003, Olanow *et al.* 2003). In parallel to the work on PD transplant, therapies were trialled for HD.

There is no effective treatment for HD, and the prognosis is poor for these patients. The first experimental studies of cell transplantation for HD took place nearly twenty-five years ago (Dunnett and Rosser 2007). Although controversial, in France and the United Kingdom, there have been ongoing trials with fetal primary-cell transplants in approximately fifty patients since 1995 (Rosser *et al.* 2002, Dunnett and Rosser 2007). Patients will receive transplants in an ongoing UK MRC-funded pilot study.[11] Assessment of patients' suitability for surgery is complex and under strict protocol. Cells are harvested from aborted fetuses of 8–10 weeks' gestation, from a different part of the brain from that used for PD. Tissue quality depends on the method and timing of the collection following surgical abortion (termination under ultrasound guidance is used, as is manual extraction, to prevent mashing of tissue), with careful transportation and dissection to ensure the cells are viable and in good condition for transplanting. More recently, the use of tissue from medical abortions has been used. Cells can be 'hibernated' for up to seven days prior to transplantation, which means that the recipient doesn't have to be prepared for surgery before the cells' viability and suitability have been assessed.[12] It has been argued that cell transplantation for HD has been shown to be feasible and is relatively safe, but assessment of effectiveness and agreed protocol for demonstrating reliability have yet to be refined (Dunnett and Rosser 2007).

The use of primary cells for transplantation has been shown to have some benefits (Bjorklund *et al.* 2003, Lindvall and Bjorklund 2004, Snyder and Olanow 2005) and has been seen as the 'gold standard for the degree of reconstruction and recovery that can be achieved by cell transplantation therapies' (Dunnett and Rosser 2006). However, a number of limitations have been identified, notably the logistically difficult problem of obtaining a supply of aborted fetuses of the required gestational age and quality for transplantation and in the numbers needed – often from multiple donors for each transplant. Fetal tissue is also comparatively 'dirty' (Kent 2008a, Pfeffer 2009b),[13] and limitations on the kinds of testing for safety relate to the 'window' within which cells remain viable and can be transplanted (Bjorklund *et al.* 2003, Dunnett and Rosser 2006, Rosser and Dunnett 2006). In sum, these early transplant experiments did not provide the basis of a commercial model or a widespread transplantation service, but were important antecedents to continuing work in the area. Most recently, there have been proposals to re-evaluate these earlier studies of transplantation for PD and to recommence a multi-centre clinical trial using fetal tissue from medical abortions.[14] Although the ability to extract and expand SCs offers a potential alternative, the technical difficulties of this have yet to be overcome, and a group of international neuroscientists have been brought together by Roger Barker to develop new protocols for this experimental therapy, in the belief that more can be learnt from these primary transplants and a review of the earlier studies.[15]

As an illustration of the social dynamics of innovation, this brief overview of fetal-tissue transplantation reveals a number of complexities. First, it illustrates some of the difficulties associated with experimental tissue and cell transplants: of sourcing biological material – in this case co-ordinating the collection of aborted fetal material – and its preparation with surgery in the neurology clinic. Co-operation between the abortion clinic, the laboratory and the neurosurgeon is critically important and rests on relatively fragile local arrangements (Pfeffer and Kent 2007, Pfeffer 2009b). As the scientists said above, this collection process could not be 'industrialized'. Therefore, standardization and ensuring the quality of cells are necessary if safety requirements are to be met and clinical efficacy is to be demonstrated. Second, an experimental surgical procedure requires close collaboration between clinical and scientific communities in order to develop new clinical protocol to construct measures for assessing patient outcomes and evaluating results. However, although these early transplant studies were seen to demonstrate functional improvements in patients, there has been tension between needing to understand better how the transplants worked before further trials can be conducted and accepting that functional improvements are a sufficient basis for developing a treatment. Moreover, one neuroscientist explained that animal studies indicate that grafting in cells stimulates the brain and halts degeneration, and that the SC work was moving away from a transplant model to a pharmaceutical approach and towards the idea of a 'stem cell medicine' and a model of regeneration:

A: So, we're not replacing the cells but somehow we're influencing what's left and getting that to perform better.

Q: So it's a sort of different concept of regeneration in a way?

A: It's a completely different concept, yeah. And my guess is, at this moment in time, in most of the disorders where we're slightly rewriting the stem cell theory in a sense [. . .] we're gonna have to rethink what stem cells do. This is important for a whole load of reasons; one is that it says that there really might be better ways of doing this than stem cells, right. So in other words, if the cells are coming along and secreting something or putting out some growth factor or something, and that's having effect, well then maybe you can find out . . .

Q: What's the something?

A: . . . what the factor is and, like I say, you could bottle it right, and you wouldn't need the stem cells and all the ethical problems go away as do a lot of other problems [. . .] But the other thing is it deals with this issue that we started off talking about, namely that some people when they looked at this field, and I'm going back a few years, thought that the whole concept was silly, that you know, you can't re-wire the brain, you can't put cells back because it's just too difficult a task. Well, what this says is well, perhaps you don't need to, perhaps that's not what the cells are doing after all you know, you're right it was too big a task but that's not actually what's going on. (SS5)

In this respect, he saw links between science, industry and the state as critically important to get new cell therapies off the ground:

A: Well, I think the cells we're producing are an off-the-shelf product. My feeling is that . . . perhaps it reflects my background but my feeling is that if you're going to have a therapy that treats anything but a tiny proportion of patients, then you're looking to the pharmaceutical industry. I mean one could argue about whether that's how it should be and that you know, governments whatever, should take a stronger role, but in the here and now that's the way it is. They're the only organizations with the infrastructure and the know-how to deliver you know, mass medicines, let's put it that way. So if we're going to treat more than a handful of people with stroke, it's going to be through the pharmaceutical industry. (SS5)

He went on to describe the similarities between cell-therapy biological *products* and medicinal products at a time when regulators were hotly debating the most appropriate regulatory framework (see Chapters 3 and 4). He argued that a medicinal approach to cell-therapy and product development was most appropriate and contrasted it with the 'developmental' biological perspective of those working with hESC, which he saw as having a very different history, but he suggested, 'there's been this convergence on this concept of regenerative medicine' (see Cooper 2005). In this view, 'regenerative medicine' is a coming together of different strands in the life sciences and the culmination of work in the field of developmental and cell biology, but also transplantation medicine.

Conclusion

What I have shown in this chapter is that the production of new scientific knowledge in the life sciences can be understood in terms of the links between the work of scientists in universities, clinicians, industry and the state. Interventions such as the UKSCB and ISCI initiative, although directed towards the standardization of scientific knowledge and stabilization of scientific objects such as hESCs, are state sponsored. The UK government has strongly supported these initiatives as a means of promoting UK Science Plc.[16] The drive to commercialize SC technologies is exemplified here by the activities of ReNeuron, a United Kingdom-based spin-out company that initially sought regulatory approval for trialling its stroke therapy in the United States, but first gained approval in the United Kingdom. The investment of public and private money to develop SC technologies and the establishment of new public–private partnerships mark the relationship between industry and the state in this biotechnology sector in the United Kingdom. In parallel, the relationships between clinicians and university-based scientists were seen as central to the fetal-tissue experiments for the treatment of PD and HD. This work, driven by a clinical agenda, was small scale and reliant on close personal relationships within a relatively small network of advocates, but it was seen as producing important knowledge in the field. However, rather than being seen as potentially leading to wider availability of such treatment, it was seen as a precursor to further basic scientific work on the biology of SCs and cell-based therapies for the treatment of neurodegeneration.

Let us, therefore, return to the question posed at the beginning of this chapter – does regenerative medicine represent a new paradigm, a new approach to the treatment of disease and illness, new ways of seeing and being in the world? The biological and social are interconnected; technical innovation is tied to social networks, political and regulatory institutions. We have seen how new biological concepts such as plasticity, potency and immortality are historically and socially contingent, and that SC science is a politically charged endeavour. Attempts to legitimate and standardize the use of embryos as a source of SCs have been shaped by political concerns to seek advantage in an emerging field and to secure investments that will create a market for SC products and therapies. The prospects for regenerative medicine and SC technologies have been linked to the achievements of the past, to 'success stories' of bone-marrow, skin and cornea transplants. Bone-marrow transplantation in particular has been reconfigured as a highly successful SC therapy (Brown 2003, Brown and Kraft 2006, Brown *et al.* 2006b, Kraft 2008, 2009). Certainly, we have seen that reports of discovery and new breakthroughs in the laboratory are central to the notion of a 'promissory stem cell science'. Supporters frequently deploy a 'rhetoric of hope' (Mulkay 1997), and expectations about future therapies are high. The possibility of developing new types of 'stem cell medicine' on an industrial scale is linked to commercialization and technology transfer, and yet the evidence of success is so far patchy and highly provisional.

Such an approach would go beyond a simple 'replacement' or transplant model and has, at least in terms of a model of production, been compared to the industrialized pharmaceutical sector. However, the biological characteristics of cell- or tissue-based therapies recapitulate the notion that some kind of bodily exchange would be taking place. The containment of SCs within a pharmaceutical frame has been problematic.

How, therefore, does this help us to understand the news in November 2007 that new induced pluripotent SCs had been derived from somatic cells? Well, this 'breakthrough' was seen as radically undermining earlier ideas about 'pluripotency' as a characteristic only of hESCs, because the lines were derived from fibroblasts. Moreover, they were fibroblasts from diverse *sources* – a young woman's face, an older man's joint, a newborn-boy's penis and a 16-week, female aborted fetus. Contrary to attempts to standardize procedures for both collecting embryos and deriving SCs, these experiments seemed to propose that heterogeneity is both meaningful and productive. This news revealed how the stabilization of scientific categories and concepts has not yet been achieved. The news was also considered politically favourable, as it was reported as resolving the ethical dilemmas of using human embryos as a source material, even though the Thomson laboratory had used fetal fibroblasts (a cell line known as IMR-90),[17] derived from an aborted fetus, which might be considered politically sensitive in some fora (Kent 2008b). The speed with which ReNeuron wanted to associate itself with the news reflected the way in which links between bench science and clinical therapies are sometimes made. ReNeuron, while still awaiting regulatory approval to trial its therapy, sought to validate and legitimate its technological approach and to stake a claim on the market prospects for a therapeutic *product* based on it. While the scientists' efforts were directed towards producing new *data* to support claims that somatic cells could become pluripotent, the company's efforts were directed towards creating a market for its product and ensuring that investors remained buoyant about the company's development strategy. However, the regulatory obstacles facing such companies should not be underestimated, as the next chapter makes clear.

3 Regulation and governance of tissue- and cell-based therapies in Europe
Ethical controversy and the politics of risk

Introduction

As I described at the beginning of Chapter 1, the sourcing of human biological materials for use in research and health treatments raises issues relating to the circumstances of the procurement (for example. donor selection, consent), the processing of those materials – whether as whole organs or divided into smaller parts (tissues and cells) – quality and safety (free from contamination), storage of the materials (for example in a tissue bank) and distribution. Additionally, the application of industrial processes and commercialization of tissue-based technologies raise controversial ethical issues relating to the 'commodification' and exploitation of human tissue. These complex exchange relationships within 'tissue economies' have been a focus of increased regulatory activity at national and international levels (Faulkner *et al.* 2006, 2008, Kent *et al.* 2006b). Biotechnology innovation has created new possibilities for the use of human tissues and cells and is associated with the emergence of new regulatory institutions and instruments (Brown *et al.* 2006a). This chapter examines the relationship between regulation and innovation and how new regulatory institutions and instruments have developed in the EU that govern the production and use of tissue- and cell-based therapies.

The incident of stolen bone material being used in the NHS and the adverse effects of implanting 'research-grade' cord blood cells into a British patient pointed to the global context of regulatory systems and introduced, briefly, the work of three regulatory bodies, the US FDA, the UK MHRA and UK HTA (see Chapter 4). My discussion here explores regulatory policy-making in Europe since the late 1990s and the shaping of legislation and governance processes. The dynamics of policy-making can be characterized by tension between national traditions and values and attempts to construct Europe-wide regulation that harmonizes member states' laws while, under specified circumstances, permitting national variations (Faulkner *et al.* 2006, Kent *et al.* 2006b). This multi-layered process merits investigation, and I highlight here both the specific policy 'problems' associated with regulating the collection and use of human biological materials and broader weaknesses in the European governance system.

To begin, I describe the key controversy surrounding the commodification of the human body and how this shaped debates around the 1998 EC Biotechnology Directive (Directive 98/44/EU) and the patentability of human life forms. Then, drawing on interview data, I examine the political conflict that led to what has been referred to as a 'bifurcated approach' to risk governance of human materials (Farrell 2009). I see this as related to the conflict between the political interests and ethical standpoints of the tissue-banking community on the one hand and industry on the other. We see here how these stakeholders sought to influence and shape the 2004 EU Tissue and Cells Directive (EUTCD) and the 2007 EU Regulation on Advanced Therapy Medicinal Products (Brown *et al.* 2006a, Faulkner *et al.* 2006, Faulkner 2009a).

Ethical controversy and the principle of subsidiarity

Ethical concerns relating to developments in biotechnology have shaped national and international debate in diverse ways. These debates have, to an extent, intensified with the emergence of newer genetic and SC technologies. At their centre are concerns about the commodification of the body, the ethical acceptability of patenting human forms of life and questions about whether the human body and its parts can be owned, and the implications of commercial exploitation or use of body parts. Many discussions begin with the famous legal case in the United States of John Moore, in the 1980s, who, while receiving treatment for leukaemia, had his spleen removed, and, unknown to him, his treating physician used it to create and patent a cell line that had commercial value. This case highlighted the problem of whether Moore could claim that the tissues and the cell line derived from them belonged to him: whether he had a property claim to them. In the end, although it was accepted that he had been wronged, in so far as the materials had been collected without his consent, the court found that he had no property rights to the parts taken from him or the cell line, and therefore there was no validity to his claim to a share of the profits that arose from the use of the cells (see, for example, Dickenson 2007). The court's decision upheld a particular view of bodies and the inalienability of body parts.

Concern about commodification is commonly tied to the idea that these practices arise as a result of a capitalist system of exchange that, in treating bodies as objects rather than subjects, undermines notions of personhood and what it means to be human. In other words, capitalism can be understood as resting on a moral economy where there is a separation of the realm of persons (subjects) from that of commodities (objects). However, anthropologist Klaus Hoeyer suggests that new types of property relation are being created through the patenting of biological life, and that patenting disputes such as those surrounding the patentability of hESCs co-construct them as persons and commodities (Hoeyer 2007, Hoeyer *et al.* 2009). He argues that, as 'human boundary objects', exchanges of tissues, cells and body parts are embroiled in a dynamic system whereby policymakers, companies, donors and others negotiate the boundaries between persons and things with certain effects. By tracing the historical roots of patent law, he shows how 'patents operate as a form of power granting particular entitlements to selected actors' (Hoeyer 2007:335) and that 'the history of human gene patenting can also be read as a story about demarcating the physical human body – as a locus for the person – from trade' (ibid.:333). Therefore, although there are frequently assertions that tissue donors should not be paid for donating their tissues and cannot claim property rights in the body, patent claims in the United States (and elsewhere) rest on the test of inventing a *process* or adding human labour to a naturally occurring biological substance. Therefore, 'cell lines and genes could be patented because the patent related only to the enhancement, not to the human person per se, not to the material in "its natural state"' (ibid.:338). It is therefore the *labour* of those who create a cell line, rather than the donor of the source material, that is acknowledged as the basis of a patent and property claim. Giving the donor property rights would mean acknowledging the contribution of their body. In this way, Hoeyer argues, the inalienability of the body is maintained, while commercial exploitation of the products derived from it becomes possible, and patent law attempts to provide a solution to the problems of managing the boundaries between persons and things – property relations. Moreover, he suggests that, ironically, those who are critical of the commodification of bodies in the new bioeconomy can, according to this analysis, be seen as reinforcing the same moral order that underpins the capitalist system they seek to criticize.

Controversies around the patentability of human life are culturally shaped, and it is widely recognized that the moral issues of whether or not human biological materials should be patentable cannot be resolved by legislative measures at the international level (European Group on Ethics in Science and New Technologies 2000, 2002). In her discussion of the forging of a political identity in relation to biotechnology, Jasanoff says:

> I have suggested throughout that the politics of biotechnology at the EU was subordinated in key respects to that of the member states. Basic questions about the acceptability of biotechnology's products and the allowable forms of debate concerning them remained national in character.
> (Jasanoff 2005:280)

In Europe, there have been attempts to harmonize patent law and reach consensus about how the tests of invention and novelty are applied, but, in relation to patent claims for cell lines derived from human embryos, disputes have been particularly hard to resolve (Plomer 2004). In order to accommodate and recognize cultural differences, the much-debated EU Biotechnology Directive 98/44/EC identified certain exclusions to patentability:

> (37) Whereas the principle whereby inventions must be excluded from patentability where their commercial exploitation offends against *ordre public* or morality must also be stressed by this Directive;

And:

> (38) Whereas the operative part of this Directive should also include an illustrative list of inventions excluded from patentability so as to provide national courts and patent offices with a general guide to interpreting the reference to *ordre public* and morality; whereas this list obviously cannot presume to be exhaustive; whereas processes, the use of which offend against human dignity, such as processes to produce chimeras from germ cells or totipotent cells of humans and animals, are obviously also excluded from the patentability.

And Article 6 states:

1   Inventions shall be considered unpatentable where their commercial exploitation would be contrary to *ordre public* or morality; however, exploitation shall not be deemed to be so contrary merely because it is prohibited by law or regulation.
2   On the basis of paragraph 1, the following, in particular, shall be considered unpatentable:
    (a) processes for cloning human beings;
    (b) processes for modifying the germ line genetic identity of human beings;
    (c) uses of human embryos for industrial or commercial purposes.

National differences in views about whether or not there is a potential threat posed to morality (*ordre public*) by patenting laws and decisions at the international level have subsequently been seen as hampering the work of the European Patent Office (EPO). Whether patents can be issued is considered critical to the commercialization of SC therapies, the promotion of trade in the EU and the competitiveness of the EU in world markets and was hotly debated at a meeting on *commercialization and patentability* hosted by the EU Framework-funded EUROSTEMCELL project in Lund, in March 2006, which I attended.[1] Protection of intellectual property rights (IPR) and patent protection are central planks of contemporary biosciences and biotechnology innovation and were seen by many workshop participants as of critical importance in promoting European competitiveness within the global bioeconomy (see also Cooper 2008, Waldby and Salter 2008). There are national patent laws, but the EPO can issue patents that apply in more than one country. Therefore, it is possible to apply for a patent separately in different countries, or apply for a Europe-wide patent. At this meeting of SC scientists, ethicists, legal and other academics, and patent officers, there was disagreement about whether or not patent law could be used to resolve morality issues, and the EPO was heavily criticized for attempting to decide morality issues while interpreting the law. The EPO officers, for their part, asserted that its role was not to determine what the law was, only to interpret law passed by the EU Parliament. At that time, they had suspended decisions on all applications for hESC cell lines. It was also asserted by leading scientist Anne McLaren that SC lines are 'cultural artefacts', that these are *products*, and are 'not found in nature' and therefore should be patentable. Dispute focused around what was meant by 'uses of human embryos for industrial or commercial purposes' and the implications of seeing hESC lines as products, that is, the results of manipulation and processes that could be patentable. Crucial to the argument was whether the history of such a product, originating from the human body, meant it was unpatentable and, indeed, whether an embryo is a body part. Some scientists argued that the blastocyst (pre-implantation embryo) is not a body part, because it is genetically distinctive and has not been attached to a body, but is a human organism created *in vitro* (fieldnotes, March 2006). Another legal academic argued that what was being neglected in such debates was how patent law was intended to create monopolies and how the morality of commercialization itself was monitored. According to him, there was a need to distinguish between the morality of the scientific practices and the morality of granting a patent for discoveries and, therefore, to recognize that the former fell outside the scope of the patent office or patenting process (Laurie 2004). Challenges to the 'Edinburgh patent' for a method of somatic-cell nuclear transfer highlighted the weaknesses of the EU patent system and 'the failure of international standardization in the governance domain of hESC patenting' (Waldby and Salter 2008:16).[2] More recently, the Hinxton Group, an international consortium on SCs, ethics and law, argued for greater co-operation in the field, with improved methods for data sharing and measures to overcome the negative effects of 'proprietary structures' (IPR) because, according to them:

> While the proprietary dilemmas currently faced in SC science confound many if not all areas of cutting edge life science, they are especially pronounced in the field of stem cell research. First, the tree-like shape of cellular differentiation makes the field especially prone to IPR holdings that can function as tollbooths to broad areas of work, creating a drag on investment and slowing down basic research. Second, the consequences of such slowing are especially severe in the stem cell field, where novel cell lines, reagents and related technologies function as platforms for broad areas of follow-on work. Third, the competition to stake out aggressive patent positions is accentuated in the current context of competitive national innovation policies featuring stem cell science.
>
> (The Hinxton Group 2010)

Such consensus statements clearly identify the nature of the problems arising from patent laws and efforts to commercialize SC technologies, but the political will needed to effect such changes clearly implies that national governments must set aside competitive interests for the sake of a greater 'public good'.

It has also been argued that EU patent law and the politics of gene patenting are gendered, and that heated debate around the patentability of gene sequences during the 1990s can be understood as related to fear of the feminization of bodies. According to this view, patent law, in emphasizing invention and the inventive process and contrasting this to that which appears 'in nature' or 'dumb matter', invokes a set of deep-seated patriarchal values that 'chimes with a powerful, highly gendered, cultural world view in which an implicitly male guiding force fertilizes a passive, feminized "nature"' (Dickenson 2007:122). In addition, as men can only contribute genetic identity to reproduction (not gestational parenthood), and therefore genetic identity is highly valued, the patenting of genetic sequences is seen as particularly threatening, because it challenges genetic essentialism:

> Genetic essentialism serves a patriarchal purpose and reflects profoundly patriarchal values. If the genetic is the true source of human identity, and if the genetic is reduced to the level of a commodified object through patenting, then human identity is reduced to the same level. That is why genetic patenting evokes greater fears than commodification of human ova or cord blood, which are seen as instances of 'mere' matter, feminized flesh, even waste. If this argument is correct, then the fear of widespread genetic patenting is actually a fear of being reduced to 'mere' matter, and also to female status. Biopatenting does not actually reduce all bodies to female status, but it is feared because it appears to. We do all have 'feminized' bodies now, however, to the extent that all bodies are the site of these insidious fears about objectification and commodification.
>
> (Dickenson 2007:123)

What Dickenson argues is that we should think of property rights as a bundle of rights (following the Lockean view), and different groups of tissue donors should be entitled to different rights within this bundle according to their contributions. For her, donation of reproductive tissues – ova and cord blood – should lead to greater entitlements than, say, genetic material (derived from blood, urine) stored in biobanks – not because they have greater inherent value, but because greater effort (labour) is involved in the donation. Women donating ova for stem cell science endure ovarian stimulation and extraction; cord-blood collection (arguably) affects the woman's experience and the management of childbirth. Furthermore, in her analysis, cultural differences explain the different positions nation states have taken on the commodification of the body and commercial use of human tissues and cells.

In her study of the French resistance to commodification of the body, the relationship between citizens and the state is central to understanding public policy and national bioethics opinions. Rather than a view of citizens as autonomous and able to give informed consent to donate (gift) tissues, in the French case they are subject to the state, and their bodies are, in certain respects, owned by the state (see also Rabinow 1999), 'the object of the nation's *patrimoine*' (ibid.:152) (wealth) placing limitations on the types of gift they may make. An altruistic gift of tissue can be voluntarily given as an expression of social solidarity, but French citizens:

> are subjects insofar as they are members of the *patrie* and share in its *patrimoine*, but they lack full control over their bodies insofar as those bodies are part of the *patrimoine*. They are in fact both subjects and objects.
>
> (Dickenson 2007:152)

As such, they are forbidden to sell their bodies or body parts. In practice, this has restricted the abilities of individuals to claim property in their own bodies, but, once gifted, tissues and cells can be used commercially. Hence, France (and especially its National Bioethics Council, CCNE) strongly resisted the marketization and commodification of bodies by protesting about the Biotechnology Directive, while, at the same time, it was willing to import embryonic SC lines from other countries, thus preserving its own national heritage. However, since 2004, a relaxation of the principles and concessions to commodification have been in evidence (Dickenson 2007) and were played out in debate around proposals to develop new regulation in the tissue-banking and medicinal-products sectors discussed below.

National culture and values therefore impacted both on the difficulties in achieving political consensus around the 1998 Biotechnology Directive and subsequent interpretation of it in relation to SC research, which had not become established science at that time. These disputes also led to a moratorium on funding by the EU Research Directorate on SC derivation and diversity of opinion and policies across member states relating to the use of human embryos (just one source of SCs) in such research (Waldby and Salter 2008, Salter 2006a, 2007a, 2007b, European Group on Ethics in Science and New Technologies 1997, 1998a, 2007).

It is not my intention here to elaborate in detail the political and ethico-legal disputes relating to EU patent law and the implementation and operationalization of the Biotechnology Directive or the challenges some member states (e.g. the Netherlands, Italy, France) made to the legality of the directive and their resistance to implementing it, but this is part of the context of a more detailed discussion of my investigation of the building of two other interrelated regulatory regimes in Europe. This context also relates to the international concerns around blood safety that led to the Blood Directive (Directive 2001/83/EC) and more recent discussion about how to promote safety standards for organ exchanges (Farrell 2009). In the arena of European regulation, the difficulties of resolving ethical disputes have led to the invoking of the principle of 'ethical subsidiarity' in order to accommodate national differences (Faulkner *et al.* 2006) and what has been called a 'bifurcated' approach to risk governance (Farrell 2009). The role of bioethics and bodies such as the European Group on Ethics (EGE) in Science and New Technologies in the policy process is also of interest and has itself been the focus of a study highlighting its role as promoting the acceptance of biotechnology and mediating ethical conflict. Indeed, it has been suggested that, 'it is probably not too much to say that, without the ethical stamp of approval from the GAEIB/EGE, Directive 98/44 might never have been adopted' (Busby *et al.* 2008:814).

Banking communities

Regulation can be seen as having a number of functions, including managing risk, facilitating trade and engendering public trust (Faulkner *et al.* 2006). This constructs a dynamic between stakeholders, shapes the policy process and creates tension. These tensions between competing stakeholder interests have in turn influenced the building of a new regulatory regime for tissue- and cell-based therapies, which has been played out in the development of two new European regulatory instruments – the EUTCD (Directive 2004/23/EC; European Parliament, 31 March 2004) and the European Regulation on Advanced Therapy Medicinal Products (RATMP) (Regulation (EC) No. 1394/2007) (Kent *et al.* 2006b).[3]

Since the 1990s, the scope of the EC's regulatory activities has widened in order to promote a single market, and 'new approach' directives have sought to harmonize national legislation in relation to health-care products, especially medical devices and medicinal products (Kent and Faulkner 2002). Study of how the case for a new regulatory regime was framed and the building of TE as a new 'technological zone' has revealed a complex interplay of policy networks, competing discourses and stakeholder interests, combined with scientific and clinical uncertainties about the benefits of these new technologies, how the risks associated with them might be assessed and their efficacy (Faulkner *et al.* 2006, 2008, Kent *et al.* 2006b). There have been 'boundary disputes' between constituencies around the demarcation between tissue-banking and TE activities and associated with this the framing of newer technologies as distinct from more 'traditional tissues'. Within the EC, the Health and Consumer Protection Directorate-General (DG Sanco), under the terms of Article 152 of the Amsterdam Treaty, proposed and developed the new TCD, while the Enterprise and Industry Directorate-General, acting under Article 95, led the development of the new regulation for market approval of what became termed 'advanced therapy medicinal products' (RATMP). As with debates around the Biotechnology Directive, ethical concerns about the commodification of the human body and the commercialization of tissue- and cell-based therapies shaped debate and policy-making.

In 1994, the Council of Europe recommended that common action be taken to harmonize legislation in member states on human tissue banks,[4] and, in 1998, the EGE drew attention to ethical concerns relating to tissue banking in one of its opinions advising the Commission:

> Wherever tissues are removed from human beings, and possibly transplanted into other human beings, the activities involved in the collection

and use of such tissues are subject to ethical requirements intended to safeguard respect for human beings, their dignity and autonomy, and for the common good. The issue of safety is also vital, as the European Union has set itself the objective of guaranteeing each citizen a 'high level of human health protection'. This protection must extend to tissue donors and recipients, and to all health care professionals – whose work involves collecting, manipulating and using human tissues. At present we cannot be sure that the health safety of tissues is properly ensured throughout the European Union. This is due to the shortcomings of the existing national rules. As a result tissues move freely within Europe and are sometimes even imported from non-member countries, in many cases without detailed information on their origin, and in particular on the state of health of the donor.

(European Group on Ethics in Science and
New Technologies 1998b: para 1.8)

At that time, there were a number of concerns about: the safety of tissues used in transplantation; the risks of disease transmission and contamination of tissues and cells; the need to protect the integrity of the body, to protect donors and recipients and to prevent discrimination and exploitation of tissue donors; how to promote social solidarity, tissue availability and equitable access to tissues and tissue-based products; and the setting-up of a licensing system for tissue banks. Following the Council of Europe's recommendation and guidance, the EC sought to put in place the legislative framework for new standards and controls (Tatarenko 2006).[5] Between 2002 and 2004, the Commission consulted about its proposal to develop new standards, inviting stakeholder comments, and the EU Parliament convened an industry-sponsored hearing in 2003 to discuss the proposals. The hearing highlighted conflicting views about the scope of the proposed legislation, the adequacy of the process for resolving ethical differences, especially relating to the use of embryonic SCs and trading in body parts, and the concerns of industry (for-profit tissue banks) to create 'a level playing field' (see below) with non-profit tissue banks in the health services (Kent *et al.* 2006b, Faulkner *et al.* 2006).

From fieldwork carried out during this period, including interviews with different stakeholder groups and key members of the Commission, I gained insights into the framing of policy issues and the positioning of different constituencies in the policy debate.[6] For example, accreditation and oversight of tissue banking in Belgium was established in 1988, and tissue banks there had to be 'not-for-profit' to prevent commercialization of the procurement process, as one tissue banker explained in an interview in 2003:

Q: The ethical problem is still the one about the procurement not being from a hospital?
A: Yes. And a certain point of view, a very important point of view in Belgium, is the problem of medical devices. We fully agree some

tissue-engineered products can be considered as medicinal products and in the field of commercial aspects. We fully agree with this with one big exception – for the procurement. The procurement has to be a donation and with all characteristics of a donation. That would be no personal advantage, that you can accept that somebody can have reimbursement of the costs.

Q: So no personal?

A: Yes, no personal profit. No, and no personal profit. And certainly no personal profit also for persons who perform the procurement for personal profit was involved . . . And that's why we don't accept commercial (organizations) can perform such procurement.

Q: Does that also mean that commercial organizations, in the Belgian point of view, that a commercial organization cannot be . . .?

A: A commercial organization cannot be involved in the procurement anywhere. (R-EU2, tissue banker)

Thus, he emphasized the need for procurement to be free and voluntary, and tissue donation an act of social solidarity that should be separate from the commercial use.[7] He explained that, for these reasons, they had been unwilling to permit the use of the product Apligraf, which had been derived from foreskin collected in the United States by the company Organogenesis – the circumstances of procurement were considered unacceptable to the Belgian authorities. In his view, companies processing tissues and cells should be required to have agreements with not-for-profit tissue banks for the procurement of tissues and should not be able to procure the tissue for themselves. Moreover, one danger of allowing procurement to be profit generating would be a decline in the willingness of people to donate tissues:

For in Belgium it would be very difficult to introduce the profit about tissues and cells. It's very hard. It's a problem of solidarity. And in Belgium, if you introduce it would be a danger of this way to consider (tissue) a medicinal product. If in Belgium the public will know that somebody can profit about tissues and cells, and even probably about the processing also, we will see in Belgium a drop, a drastic drop in the donation. And I think this is an important challenge, because I think also the Belgian authorities if they can agree with non-profit for tissues and cells, they can agree – and I also can agree with a profit for processing in tissue engineering I think is high technology, the processing of tissue engineering has to be paid. There are people with money to invest in research, and this money is at risk. And the problem of this is to have a financial return from this research, even if this research is done about free material. The material must be free. But, in Belgium, to let Belgian people understand this difference profit from processing. Profit from this is compatible with the non-profit for donation of the material. It's a very big challenge this problem. (R-EU2)

The institutional arrangements for procuring tissues were of central concern, therefore – considerable importance was attached to how donors gave tissue voluntarily and that they were unpaid. The challenge was to regulate how such tissues subsequently entered the market or were traded as tissue and cell products. Tissue donation at the point of procurement was best seen as a non-profit, public health service. At the same time, this banker was prepared to accept that the *processing* of cells could be costly, and therefore, profitability would be important if cell products were to be made available as a treatment. For him, however, procurement needed to be viewed separately from the commercial use of tissues.

In France, since 1994, cells and tissues had been regulated through tissue-banking legislation that set standards and implemented an inspection system, but 'high technicity' products were regarded as medicinal products, and, in 2003, this tissue banker, who represented France in many discussions about the new regulations and was a key figure in the international forum in Porto (2000) and Malaga (2002) leading to the drafting of the proposed directive, suggested that the concept of 'tissue engineering' did not exist and couldn't easily be defined:

Q: [. . .]But could you give a brief outline of the current situation for tissue-engineered products in France?
A: Tissue engineering does not exist.
Q: Officially?
A: The terminology doesn't exist. We regulate tissues on one part and cells on the other part. (I keep organs out of the discussion, right.) To be able to use tissues, these tissues have to be authorized and products in an authorized tissue bank. There is a decree, there are standards and inspection. Concerning cells, it's the same approach but different decree, but the same year, but the approach is the same. You have to have an authorized cell bank, processing organization, and authorized products. Well, we used to speak about products for cells and process authorization for tissues, but it's very un-useful difference I mean.

And:

A: [. . .] Actually nobody knows really where is the difference between cell and cell therapy. But it doesn't matter because it's the same authorisation. [. . .] The simple tissues create finally the same goal as the most complicated cell therapy.

Hence, cell therapies were regarded as transplants and fell within the purview of the tissue-banking regulations in France at that time. However, the expansion of cells to create 'batches of product' was seen as distinctive:

A: There is a concern depending if you create batches of product. In most cases today you have no batch. It's one for one, or one for few. So it's completely different if you expand cells, create a massive working cell bank and so on. But anyway, if you create batches for hundreds or thousands of patients. For me, if you create batches it's medicinal product. It's no more transplantation. (A-EU4, tissue banker; emphasis added)

In this view, therefore, autologous cell therapies such as *Epicel* and *Carticel* (cultured cells) could be regulated under the cell-banking regime and were not regarded in France as TE products but simply as expanded cell products, which were permitted for use as part of a clinical trial. Where there is batch processing for allogeneic products, then medicinal-product regulation (MPD) regulation was seen to apply. What he called the 'new frontier' of engineered tissue was best regulated via the pharmaceutical, medicinal-product, regulation. Yet confusingly, at the time, there was no consensus about what constituted a tissue-engineered product, and the sociopolitical process of regulatory innovation was itself tied to the building of such agreements and ordering of this new sociotechnical field. Processes of 'purification' and 'partitioning' could be identified, as stakeholders attempted to define a clear terrain for TE and construct it as a 'technological zone' (Faulkner *et al.* 2006).

The two respondents above typify the position of the tissue-bank sector as occupying a moral high ground by providing a public service, mostly on a not-for-profit or cost-recovery basis, for sourcing tissue ethically and protecting the integrity of the body and distributing it according to need rather than for commercial gain. For them, the existing tissue-banking arrangements could accommodate expanded (autologous) cell-therapy products. However, high-tech products that required higher degrees of processing, 'manipulation' or 'engineering' were considered more like medicinal products than 'traditional tissues' and therefore should be regulated as such. In contrast, industry, motivated by commercial interests and delivering a health-care product, was unwilling to concede the high ground to the tissue-banking sector, or indeed to accept a relationship of dependence on tissue banks for the sourcing of tissues. Industry bodies such as EUCOMED and EuropaBio therefore lobbied for the inclusion of commercial establishments as eligible for licensing to source and store tissues,[8] and their members could be seen as undermining the higher moral claims being made by tissue bankers in these kinds of remark:

In Germany, autologous tissue, in our experience anyway, has a very bizarre and strange classification. In discussing how the product works and what the product is with Bfarm, they suggested that autologous chondrocyte implantation should be classified not as a medicinal product, although it should be covered by the AMG or drug law. It should be classified as what's called an unfinished drug. The drug itself would be the cartilage plug presumably, but the unfinished part are the chondrocytes which you implant at that time with surgery. Now because it's an unfinished drug there are no approval processes. So although you have to follow in spirit the rules that are outlined in the drug law you can't effectively get product approval. So you are stuffed when it comes to looking at reimbursement. So that makes things in Germany from a commercial perspective, it makes things a lot easier for these very small companies. And if you think of autologous tissue, if you were to be perfectly honest and frank, it's not rocket science. It's just a question of taking a biopsy, isolating the cells, growing the cells *in vitro* and re-implanting them. So it's all pretty much standard techniques, techniques which can be engineered within a hospital tissue bank for example. So in Germany a number of local groups have emerged within hospitals which do these sorts of products for surgeons, just covering their expenses and covering the holidays to Barbados with the tissue bankers. Oh God I shouldn't have said that 'cos it's on tape. But you know what I mean. They sort of work on a very local basis, they may cover three or four hospitals in a local district, they speak with the local *Länder* who have the ultimate authority because Bfarm doesn't give centralized approval, so things have evolved down to the *Länder*. [. . .] So there's a lot of political interest as well as scientific interest in Germany but it makes it difficult for a multinational company to get involved in that system. (M-EU9, regulatory affairs manager, multinational company)

He contrasted the situation in Germany with Spain, where, at that time, tissue-engineered products could be released on to the market, but only through a tissue bank, which charged high fees to companies to do this:

All they do is receive that piece of paper they charge a lot of money for it. And in charging a lot of money for it, it means that you have to add a premium to your product which is already expensive. So again, why should a hospital pay a premium price plus 20 per cent for a product which they can manufacture using their own little tissue bank? Absolutely no reason whatsoever. So it kicks the bottom out of your commercial market from a multi-national perspective. (M-EU9)

He also described obstacles to market approval of tissue-engineered products in France. Therefore, not only were tissue bankers described as themselves motivated by making money, but the regulatory variation in different member states was an obstacle to commercialization and product licensing for a multinational company at that time. Therefore, the trade associations sought to amend the proposals to ensure a level playing field was created between tissue banks and industry:

> In fact, if I could sort of summarize a whole group of amendments in a simple way. I think most aimed at ensuring that somehow the industry-based people and the tissue bank people are all treated like on a level playing field because *industry may act as a tissue bank* and frequently does, they treat many bank tissues, tissue banks, are traditionally acted as tissue banks by hospitals with tissues that they need. But tissue banks could just as easily put the tissue products on the market and we have no problem with that, provided that the rules are same for everybody. So most of our amendments are going in this direction to ensure a level playing field for all the actors involved. (M-EU2, EUCOMED member; emphasis added)

Tissue banks' practices of processing and storing cells were contrasted with the established GMP, which guided the activities of batch production and multinational commercial processing of autologous (or other) cells. Tissue banks were perceived by others as higher-risk environments and perhaps unwilling to embrace new regulation that could impact on their practices:

> Our main thrust within the company is to try and develop this new regulation with the regulators and to such a level which produces a more level playing field for all concerned. It's understandable that tissue banks manufacture products but at the same time it's also understandable that, without proper controls on safety, manufacture, et cetera, et cetera, they could potentially be introducing products to patients that are infected, that there's no follow up, that there's no control over the way the surgeon uses them and so on and so on. So you have to develop a system which ultimately defends patient safety. That's what regulation does. Regulation is there to make sure that the patient gets product which is manufactured properly, which is safe and effective and we in industry recognise that you need that for everybody. I remember once I went to the UK Medical Devices Agency and gave them a presentation on products and they came to see us, they looked at our processes and procedures and stuff. And they said to us at the end well, to be perfectly honest we're not really worried what [your company] does because we know that [your company] has the processes in place, there's the people following up, all that sort of stuff. What we're worried about is the fact that you get Joe Bloggs down the road who can do exactly the same as you in his garage in West Sussex, you know? And those people can't possibly maintain a level of safety that big companies can. So, understandably I think that the tissue banking community is very worried about this developing legislation. Now the way

that we input into that is firstly by trade associations but also by building up personal relationships with the people in the Commission themselves. (M-EU9)

The industry lobby therefore maintained that, (a) industry should be allowed to act as a tissue bank, and (b) that tissue banks should be required to comply with the same stringent controls as industry when processing cells and cell products. This had a significant effect on the parliamentary debate in Europe and the shaping of the TCD, which eventually introduced the concept of a 'tissue establishment' to encompass both not-for-profit tissue banks and commercial banks.

Commercialization of product development in order to make products publicly and widely available, that is, to achieve distributive justice through the market, was itself perceived by industry stakeholders as ethically desirable. Interestingly, patients' interests were aligned with those of industry in so far as they welcomed advances and the prospect of treatments for rare and untreatable disorders, while Christian and Catholic activists, such as Christian Action Research and Education (CARE) who spoke at the EU Parliament hearing on the proposals in 2003, raised other ethical concerns relating to SC research and opposed the destruction of embryos or use of fetal tissue as part of the 'cannibalisation of one human individual to provide "spare parts" which will improve the life chances, or even be essential to the survival of another human individual or individuals'.[9] Dr Peter Liese, a Christian Democrat MEP and rapporteur for the proposals, was a key figure in leading the debate and keeping ethical concerns at the forefront of discussion.

The Amsterdam Treaty provided the legal context for the Commission responsibilities with regard to public health, as a Commission official explained:

Well the point is that in Europe it is not so easy to state the boundaries between what is the concept of transplantation as such and what is the concept of tissue engineering. Such is very difficult to define precisely: what is a tissue engineering product, in a way. So of course, or we can discuss this later, there are some links and some boundaries between what is transplantation and what could be tissue engineering, that are not completely clear for one side. And for the other side it is the policy of the Commission that any source of human material, any cells, tissues and organs will be used or will be applied in human and that doesn't matter if it is going to be applied for transplantation, for medicinal products or for medical devices or for tissue engineering products; it has to be covered in there, in the first steps with the donation procurement and testing by a public health perspective. And that's why we are trying to work with, and have agreed with DG Enterprise and the Council and the Parliament. (A-EU5, Commission official)

Given the different jurisdictions of the two directorates, DG Sanco (protect-ing public health) and DG Enterprise (regulating the market), their aims and remit were to an extent divergent and in tension with each other. Indeed, there were reports of tensions between the two EC directorates, but this Commission official from DG Sanco described 'active relations' with DG Enterprise and how their conversations informed the formulation of proposals. In this area, overlaps between the two jurisdictions emerged, and the activities of the two directorates could be seen as both cutting across each other's but also as interconnecting, as discussions about the borders between the proposed TCD and proposed regulation for tissue-engineered products (TEPs) ran together and seemed at other times to run in parallel. In short, there were key boundary disputes about the scope of the two proposed directives that mapped on to the institutional interests and mandate of each directorate, as well as reflecting broader divisions between the tissue-banking communities, and those states where they were dominant, and multinational industries. These dynamics were constitutive of the shaping of TE as a 'technological zone' and the new 'regulatory order' (Brown *et al.* 2006a, Faulkner *et al.* 2006, 2008, Kent *et al.* 2006b).

In 2004, the TCD was approved. The regulation of tissue-procurement and tissue-banking practices was no longer simply a national matter and largely a process of professional self-regulation; instead, it would fall under the purview of the EU Commission, and member states would be required to report to it on progress with implementing new standards and the licensing of establish-ments and provide details of adverse incidents. Two technical directives were produced through the comitology process, which set out in more detail the requirements for procurement, donation and testing, and processing, storage, preservation or distribution.[10] Conflicts between the tissue-banking sector and industry, underpinned by tensions between a public-service ethic and commercial interests, shaped the outcome. The framework set standards that sought to align and license unstandardized tissue-banking practices, protect the dignity of donors and recipients, promote public health and safety through the 'purification' of tissues, and permit commercial development of tissue- and cell-based products. Industry and for-profit organizations secured the ability to become accredited 'tissue establishments' – a term introduced to denote a wider community of organizations that could be accredited than those already designated tissue banks.[11] Industry also secured limits on the impact of the directive in relation to the manufacturing, distribution and placing on the market of tissue- and cell-based products that would be covered by other Community legislation (Recital 6) (the proposed RATMP). Crucially, however, the ethical concerns relating to donation and the use of specific cell types were framed as the responsibility of member states:

> This Directive shall not prevent a Member State from maintaining or introducing more stringent protective measures, provided that they comply with the provisions of the Treaty.
>
> (Article 4)

And:

> This Directive should not interfere with decisions made by Member States concerning the use or non-use of any specific type of human cells, including germ cells and embryonic stem cells. If, however, any particular use of such cells is authorized in a Member State, this Directive will require the application of all provisions necessary to protect public health, given the specific risks of these cells based on the scientific knowledge and their particular nature, and guarantee respect for fundamental rights. Moreover, this Directive should not interfere with provisions of Member states defining the legal term 'person' or 'individual'.
>
> (Recital 12)

The principle of voluntary and unpaid donation was upheld, and donation was framed in terms of the 'altruism of the donor and solidarity between the donor and recipient', and member states were 'urged to take steps to encourage a strong public and non-profit sector involvement in the provision of tissue and cell application services and the related research and development' (Recital 18). However, 'donors may receive compensation, which is strictly limited to making good the expenses and inconveniences related to the donation' (Article 12). Consent, which was seen as a guiding ethical principle, became a matter for national guidance and legislation, but information given to potential donors was expected to cover areas specified in the directive. Procurement of tissues would have to be licensed, but could be carried out by an organization separate from the accredited tissue establishment through a third-party agreement. Harmonization of European law in this field was not possible as, under the Charter of Fundamental Rights of the EU and the Oviedo Convention, there is no provision for harmonization, and member states cannot be prevented from introducing more stringent legislation (Recital 22).

Therefore, the TCD could be seen as achieving at least three things. First, in policy terms, it marked a process of convergence between two sectors – although originally intended to regulate the procurement, storage, distribution and banking of 'traditional tissues' (for example, skin, bone, cornea), the sourcing, testing and storage of tissues that could subsequently be manipulated or 'engineered' were also included. Hence, tissue-bank communities and industry were both covered and could be accredited as 'tissue establishments'. However, they had disparate cultures and values that were not swept away by the legislation, although, for the tissue-banking communities in many countries, including the United Kingdom, significant changes in working practices and in the language used to describe them were to follow (see below). Second, the attempt to bracket out irresolvable conflict between members states and other constituencies about the acceptability of using some cell types – notably cells derived from pre-implantation embryos (hESC) or germ cells from aborted fetuses – created a kind of 'ethical subsidiarity' (Faulkner *et al.* 2006). The principle of subsidiarity and the constraints of Article 152 meant that decisions about which specific cells or tissues could be procured or stored would remain a national matter. Member states could introduce more stringent rules, and the directive set only the minimum standards and requirements. Third, although the principle of voluntary unpaid donation was secured, in order to avoid the commodification of the body, the separation of procurement from commercial use of the tissue and the effort to set standards in this field indirectly had the effect of *facilitating* a market in human biological materials, with a more industrialized tissue-banking sector that balances an 'ethics of efficiency in procurement with an ethics of respect for donors to be ensured through informed consent' (Hoeyer 2010a:145, 2010b). However, the placing on the market of tissue and cell products would fall under the new RATMP, as they were to become redefined as 'medicinal products'.

Industry and the regulatory state: regulating the market in human-tissue products

The debate about whether human tissue and cells could be regarded as a type of health-care product and should be regulated as such has had a long history. Health-care products have, since the early 1990s, been divided into two main groups in Europe for regulatory purposes – medicinal products and medical devices. These have been defined according to their 'mode of action'. In 1995, there was opposition to the inclusion of human tissues in the proposed Medical Devices Directive (MDD), especially from France, as this tissue banker explained:

A: So this was a starting point of a very long discussion which is still running now, how to regulate human tissues and cells in Europe. So the first point is France wanted to avoid any tissues and cells into the Medical Device Directive. We didn't want a new approach. We didn't want the fifty notified bodies to give marketing authorization, well the CE marking, under this directive. We wanted to be able to control procurement, which was not possible through the Medical Device Directive. Well everything was wrong in our opinion in this directive. Finally we obtained in '98 or something like that not to include human products in the Medical Devices Directive. But we thought it was interesting to have a common approach at the European level, to be able to exchange more easily tissues and cells, to have a comparable safety level in the different countries. And we wrote a memorandum at the Council of the European Union stating that France thought it was necessary to have a specific regulation for all tissues and cells and with some proposals. Then the discussion was officially closed at the Commission level and the Council level and then we continued working with colleagues from the different countries. And the Portuguese presidency organized a meeting in Porto in 2000. We made a survey to prepare for this meeting to know what was the existing regulation in the member states for organs and for tissues and cells. Roughly on ethical aspects it was a good level of regulation everywhere and for the safety and quality of tissues and cells there was a very low level of regulation. And clearly three or four countries only had a specific regulation for tissues and cells. So the experts in Porto were divided in two groups. The organ group said the problem is organ shortage. We have no problem with all the safety and quality processing in this activity.[12] Well, it's a very schematic statement. Let's say shortage is the most important. And for tissues and cells the answer was we need a safety and quality directive at the Union level. (A-EU4)

One issue identified here is the system for approval of devices through notified bodies, which are independent assessors who certify that devices comply with the regulation (Kent and Faulkner 2002). This had been criticized as being too self-serving and lacking rigour, but 'controlling procurement' and sourcing materials fell outside the purview of the medical device regulation, which was a 'new approach' directive focused on market authorization. Tissue banks were therefore described as strongly opposed to inclusion in medical-device regulation according to this senior UK scientist, EU policy advisor and member of the Scientific Committee on Medicinal Products and Medical Devices:

> Tissue banks had never been regulated in Europe. They were in a sense left out of the Medical Device Directive. [. . .] Tissue banks were a very powerful lobby and they had managed to keep outside of the framework of the Medical Device Directive for example. So that was one reason why human products were totally excluded from MDD at the time back in the '90s I'm told. This is hearsay that there was a strong lobby to keep tissue banks out of this sort of regulation. (A-EU6, UK scientist and EU policy advisor)

There was also a resistance to the idea of human tissues and cells as commodifiable products when the TCD was being discussed, as we've already seen and as this Commission official recounted:

A: And for us, for Sanco, we have a problem with the term products because product has connotations that we would like to avoid, even if it is a product. But you know all these, when you talk with the transplantation community, probably product could create confusion, mainly in the field of organ transplantation because they are very sensitive with these.

Q: Because it has a commercial connotation, like in commodity kind of sense?

A: Yes, but you know, there are terms that for example in organ trans-plantation are completely forbidden, like trade of organs or something like that. And the trade is linked probably with products. (A-EU5)

Tissue transplantation was linked strongly to organ transplantation, and the culture and values of tissue banks were built on exchange systems that foregrounded gift relationships and altruism. To redefine tissues as *products* was seen to undermine these values and conflict with them. What emerges here is evidence of the ways in which gift and commodity systems of exchange are intertwined (Waldby and Mitchell 2006), but also how regulatory discourses and powerful interest groups have shaped the framing of policy and legislation in this area. Industry lobbied in favour of regulation, which it saw as important in order to get new tissue- and cell-based technologies to market, and, as we've seen, this included securing the freedom to source tissues and be recognized as 'tissue establishments' under the terms of the TCD, but it was divided about whether a medicinal-product or medical-device approach was more appropriate for market approval of products. For many industry players, the latter offered a more 'light touch' approach, and there was strong resistance to the view that medicinal-product legislation could accommodate these newer types of 'tissue engineered' product, because the testing requirements and evidence base could not be considered the same as those for drugs according to these commentators. Yet, in important ways, these new tissue- and cell-based therapies disrupted the boundaries between devices and drugs and highlighted 'regulatory gaps' (Cox 2000, Faulkner *et al.* 2003). TE was itself a concept that had to be 'engineered', and tissue-engineered products could be seen as 'hybrid' or combination products associated with the construction of 'hybrid' regulatory institutions (Williams 1997, Brown *et al.* 2006a, Faulkner *et al.* 2006). Uncertainties about how to classify activities and where to draw the lines between them were rife and revealed the instabilities of the sociomaterial practices that sought to shape both the science and technology and the governance systems.

A: Yes. Well I think that discussion is still going on. I mean for me it's clear – now I think for everyone it's clear what is a cell therapy. Cell therapy means cells that are highly manipulated and you [. . .] use it like a pharmaceutical therapy. For example the melanoma vaccine using cells that can cause a reaction in the immune system, taking cells from the melanoma from the patient. That could be a good example of cell therapy or could be a typical example of gene therapy. Then for me it's clear what is a tissue-engineering product. It's clear that these products that combine biologicals, I mean tissues and cells, with synthetics and materials scaffolds, matrix [. . .] we call these artificial components. We have not had to deal only with the biological part, we have to deal with the quality and safety of the matrix and scaffold and with the consequence of the interaction between what is the synthetic with the tissues. This is clear under what I consider tissue-engineering products. And then there are a number of products for which the boundaries are not clear. For example, the autologous cultures of chondrocytes. You take chondrocytes from the patients, you culture, you expand the chondrocytes, you can create a tissue and then you implant the tissue. If you introduce in a scaffold – a matrix is a scaffold – for me it's more clear that this is a tissue-engineering product. (EU5, EC official)

Cultured skin (e.g. Apligraf) was seen by some to be 'more on the side of transplantation' than TE, but drawing the line between 'transplantation' and 'tissue engineering' was hugely problematic and contentious. A new annexe to the Medicinal Products Directive in 2003[13] brought somatic and gene therapies within the pharmaceutical frame, but its implementation was at an early stage during the consultations around a proposed directive for TE, and the absence of an agreed approach to regulating tissue- and cell-based therapies meant that, at that time, products were being treated differently in different countries – the same product might be regulated as a medicinal product, a medical device or as a tissue, or it was not regulated at all (Kent *et al.* 2006b). This 'patchwork' represented an obstacle to technology diffusion and commercialization (Faulkner *et al.* 2006). The basis for new regulation to facilitate trade and movement of these as health-care products is Article 95 of the Treaty, but national governments can prevent entry of goods on safety grounds, so that the highest standards of product safety must be ensured, as this Commission official from DG Enterprise (above) explained in an interview in 2003. He went on to say that products, once approved, could not legally be stopped at the borders of a country on *ethical* grounds (for example, if they were produced using hESCs). However, in the passage through the parliamentary process and co-decision procedure, the legal position on this was hotly disputed.

In 2006, some members of the EU Parliamentary Committee on the Environment, Public Health and Safety, in co-operation with the Committee on Legal Affairs, argued, in proposing amendments to the proposed regulation, that a number of ethical concerns needed to be addressed. Specifically, they argued that the regulation could not contravene fundamental human rights under the terms of the Oviedo Convention, and, therefore, that the creation of chimeras or the use of hybrids should be banned (amendments 21, 22). Moreover, they argued that there should be a 'ban of commercialization of the human body', that donation of human cells and tissues should be voluntary and unpaid, and that procurement should be carried out on a non-profit basis (amendment 20).[14] In addition, it was argued that member states had the legal right to prevent the use of products regarded as unethical in the individual country, and that the subsidiarity principle had to be upheld in this case. Therefore, while some took the view that the TCD had already addressed such concerns, others sought explicit inclusion of these ethical principles in the RATMP. The committee was divided, and these ethical amendments were ultimately rejected by the committee. However, three political groups in the Parliament, PES, ALDE and GUE/NGL, agreed a compromise package with the Council, to which the Slovakian rapporteur MEP Miroslave Mikolasik

(EPP-ED SK) objected in presenting his report to the Parliament in April 2007.[15] He took the view that, if accepted in the revised form, 'in practice this would mean that certain products would not have access to all markets', which he believed was illegal. This view was challenged by the German MEP Dagmar Roth-Behrendt (PES, DE), who argued that:

> Yes, Member states who want to ban the use of fetal stem cells should be allowed to do that and everybody who says that the Court of Justice would deny that and would put it under Article 95 is either not knowledgeable, and then giving the wrong impression I am afraid. If you read Article 30 of the Treaty, you are exactly sure that Article 30 says public morality is always a reason to make sure that a Member State can ban that.

Meanwhile, in the run up to the vote, Irish MEP Kathy Sinnott (IND/DEM, IE) referred to 'the trade in human body parts' in the Ukraine and a recent visit of the Ukranian prime minister to the Parliament, when:

> He asked for our help to end this form of human trafficking. Our help. Tomorrow's vote will determine whether we have any help to give. If we uphold the ban on commercialization and if we exclude human embryos from the scope of this directive, then we will be able to do something about the wealthy buyers within the EU.

In the early stages of the consultation process, some industry players – notably EuropaBio and EUCOMED members – lobbied hard for a 'third pillar' of legislation – a separate and distinctive regulatory instrument that acknowledged that TE products were neither simply devices nor drugs (Kent *et al.* 2006b). There were also strong objections to a centralized approval process through the European Medicines Agency (EMEA), as this was considered costly, time consuming and ill suited to assessing tissue-based products. Ultimately, these arguments were defeated, and it has been suggested that there was a shift from 'provenance' of tissues as an organizing principle for regulation, 'to a new principle in which the mode and locus of TE production has become dominant' (Faulkner 2009b), leading to the new RATMP adopted in December 2007. It combined tissue-engineered, somatic and gene-therapy products under one regulation for 'advanced therapies' and was largely welcomed by industry groups (Kent *et al.* 2006b, Hughes-Wilson and Mackay 2007) (but, see Trommelmans *et al.* 2007). A new Committee on Advanced Therapies (CAT) would be established to advise the Committee on Medicinal Products for Human Use (CHMP) at EMEA, representing a centralized approval process that was strongly opposed by representatives of the medical-devices sector (Faulkner 2009b).

Conclusion

In summary then, ethical concerns relating to the commodification of human tissues and trading in body parts have shaped international debate and the formation of a European regulatory framework for tissue- and cell-based therapies. In practice, this has led to an emphasis on unpaid, voluntary donation of tissues and cells and not-for-profit procurement. The tissue-banking community has been a powerful lobby group, and, whereas previously its practices were largely self-regulating and a national matter, it has, under the terms of the TCD, fallen under the scrutiny of the European regulatory state, although, in some countries, tissue bankers claimed to uphold moral principles and were unhappy about proposals that appeared to redefine human tissue and cells as health-care products. A discourse of 'production' and recognition of the industrial processing or significant manipulation of tissues and cells were perceived as undermining those principles and as a sign of creeping commercialization.

For their part, industry users of human tissues and cells sought to create a 'level playing field' with tissue banks and hospital-based laboratories by arguing that companies should be accredited as tissue establishments and not be reliant on tissue banks for sourcing or procuring tissues and cells, and that, where hospitals and tissue banks were processing and distributing tissue or cells, the same standards should apply to them as applied to industry. In relation to the TCD, this meant that it came to have a much broader scope than originally envisaged, extending beyond both conventional tissue-banking activities and the procurement and storage of 'traditional tissues'. The RATMP accepted that commercial use of human tissue and cells was legitimate where health-care products were approved for placing on the market via a centralized procedure that set standards for safety, with exemptions for hospitals for single-patient, non-standard applications (Faulkner 2009b). However, donation and procurement were to be not-for-profit and non-commercialized and would fall under the TCD even where commercial use of the material followed. Moreover, in recognition of ethical sensitivities and variations between member states, both the procurement of certain types of cell and the use of products derived from them could be prohibited at the national level, in accordance with the principle of subsidiarity. As a result, the potential harmonizing effects of the new regulatory regime appear weakened. However, in the settlement of the RATMP, there remains 'an uneasy Europeanization of human tissues and cells as tradeable technology and of engineered tissue as a technological zone embracing both manufacturing and some but not all hospital enterprise' (Faulkner 2009b:643), and 'tensions between the established industrial and regulatory *sectors* of medical devices and pharmaceuticals, and the divisions between commercial and hospital/tissue bank producers of therapies' (ibid.: 643) remain.

In important respects, the disputes relating to the TCD and the RATMP mirrored those surrounding the earlier Biotechnology Directive, where concerns about commodification were central to disputes about the patentability of genetic and SC technologies. This could be understood (following Dickenson) as a fear of the feminization of bodies and a more generalized degradation of what it means to be human. However, while these ethical issues and diverse national values and political cultures, as well as sectoral interests, impacted on the outcomes, the role of the European regulatory state in managing and regulating risk was the final arbiter in legislating for a market in human-tissue exchange and tissue and cell therapies. The next chapter examines the links between the policy debate at the international level with analysis of the politics of regulation in the United Kingdom.

4 A 'strict but permissive approach'

A case study of UK regulation of human-tissue and cell therapies

Introduction

While the developments in European regulation described in the previous chapter provide an important context for, and framing of, international regulatory policy relating to regenerative medicine and the new tissue- and cell-based therapies, national (political) culture and history can also be seen as significant, as we have seen. The United Kingdom has had a high profile in this international arena and, some have argued, has sought to lead debate. This chapter provides a closer analysis of the dynamics of policy-making at the national level and reflects on what might be the specific features of the UK approach to regulating these technologies. At the same time, I want to examine how the recent policy developments in Europe have been played out and implemented in the United Kingdom.

In May 2008, public attention was once again focused on the ethics of embryo research, the morality of creating hybrid embryos or chimeras, the links between reproductive technologies and new family forms and the rights of women to have an abortion. Parliament debated proposed changes to the 1990 Human Fertilisation and Embryology Act (HFE Act) (which also modified the 1967 Abortion Act), and the media reported heated, impassioned declarations about whether scientists were attempting to cross a line by mixing animal and human biological material; whether creating 'saviour siblings' was acceptable; if the 'need for a father' should be a consideration when assessing a woman's suitability for IVF treatment; and whether the age limit for pregnancy termination should be lowered from 24 weeks. While, elsewhere in Europe, other countries continue to debate and review policy on embryo and SC research, the United Kingdom can be seen as having one of the most liberal approaches, but based, it is claimed, on a strict regulatory regime. As an earlier report from the UKSCI put it in its advice to government on the prospects for SC research in the United Kingdom:

> Without doubt, the enabling and consistent regulatory environment is currently one of the strongest assets to UK stem cell research. Indeed, overseas researchers have been attracted to the UK precisely because of

its coherent regulatory framework covering all forms of stem cell research in the public and private sector.

(Pattison 2005:43)

How, therefore, might we assess these claims and better understand the political dynamics in the UK setting? The Pattison report extolled the virtues and strengths of the regulatory approach to SC research in the UK, claiming it was to be envied, or at least 'well regarded', by others and reflected the 'favourable ethical environment and public support' (ibid.:42) for embryonic SC research and adult SCs. It recommended that regulation of SCs is risk-based and proportionate and does not stifle the development of SC therapies.

The Pattison report presented a rather selective and somewhat naive view of UK regulation that sought to claim the moral and political high ground for the United Kingdom by making claims about the coherence and strength of regulation in a nationalistic and UK-centric way. The shaping of regulatory institutions and policies (the regulatory order) can be seen as linked to the mediation of the interests of patients, clinicians, scientists, industry and the state. In the following discussion, I explore the political histories of three regulatory bodies in the United Kingdom and suggest that these relationships have taken a slightly different form in each case. Attempts to protect public health, build public trust and confidence, manage risks and promote trade – all aims of public policy in this area – can be seen as mediating these interests in rather different ways. I want to consider how the responsibilities of the different agencies have been negotiated, contested and redefined in the shaping of a new 'regulatory order', which in turn can be linked to the construction of new 'regulatory objects'.

In research (with Naomi Pfeffer) looking at the collection and use of aborted fetal tissue in SC science, we investigated how people make distinctions between pre-implantation embryos and aborted fetuses as sources of cells, and how such distinctions shape and are shaped by regulation. Embryos and fetuses can be seen as distinct 'regulatory objects' that have been regulated separately. Whereas much of the public debate over the past few years has focused attention on the use of hESCs, we argued that, by comparison, fetal-derived cells seem to have received little public attention. We suggested that the arrangements for the regulation of the collection and use of fetal tissue in SC science were confused, lacked transparency and were increasingly out of line with good practice on seeking consent (Pfeffer and Kent 2006, 2007). Scientists themselves sometimes contrasted the bureaucratic burden of regulation relating to embryo research with the relatively 'light touch' approach to fetal-tissue use. These differences caused me to reflect on the political histories of the different regulatory institutions and to ask how we had got to where we were then, and how we understand the current institutional arrangements for regulating SC science and its ethical oversight. We need to see the emergence of 'embryos' – the term commonly used to refer to pre-implantation embryos – and 'fetal tissue' – a term used to refer to tissue from a miscarried or aborted fetus – as distinct 'regulatory objects', tied to the history of the institutions that have regulatory responsibilities for them. Moreover, more recently, with revisions to the HFE Act in 2008, we saw the creation of new regulatory objects called 'human admixed embryos', and, with the implementation of the 2007 EU Regulation, a new category of 'advanced therapy medicinal products' emerged, as we saw in the last chapter.

Instability, uncertainty and change have characterized the regulation and governance of 'traditional human tissues' (skin, bone), 'engineered tissues' and cell-based therapies. Scientific and technological changes are associated with changes in the regulations and in regulatory institutions, as we have seen already in Chapter 3 (Brown *et al.* 2006a). The boundaries between these institutions are sometimes contested and permeable (Kent *et al.* 2006b, Faulkner *et al.* 2008). In the United Kingdom, there is a specific context and history of regulatory change that has influenced the direction and implementation of EU legislation. Moreover, the challenge for SC researchers and manufacturers is negotiating a route through a regulatory maze. This chapter therefore begins by discussing the HFEA, then the HTA, the MHRA and, briefly, the Gene Therapy Advisory Committee. A proposed merger of the HTA and HFEA to form a new Regulatory Authority for Tissues and Embryos (RATE) in 2008 would have drawn embryos and fetuses into the same regulatory framework as other tissues and cells. However, such a move was hotly contested, at least by some strong supporters of what I call the HFEA model.

Regulatory institutions and regimes

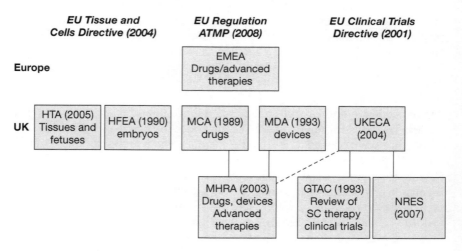

Figure 4.1 Regulatory institutions and regimes

Interestingly, in July 2010, the new UK government proposed to close down both the HTA and HFEA, and, if these proposals are adopted, this could mark the beginning of a new era for regulation in the United Kingdom, although whether this will sweep away the old politics is unclear (Department of Health 2010, Kent and ter Meulen 2011).

Progressive science and UK science policy

Since the 1980s, the Warnock report in 1984 and the passing of the HFE Act in 1990, the 'embryo research debate' has been a focus of study for sociologists, ethicists, political scientists and legal scholars (see Appendix 1). According to Michael Mulkay, 'the examination of the rights and wrongs of embryo research in Britain during the 1980s was highly unusual in the degree to which it subjected a particular branch of scientific inquiry to sustained, collective appraisal' (Mulkay 1997:2). Moreover, in his analysis, the struggle over embryo research can only be understood in terms of its social and cultural origins, and a number of 'cultural' tensions have continued since the 1990 legislation was put in place. A key feature of policy-making at that time was the level of public involvement (and protest) in the national debate. This has continued to be the case, as the actions of the regulatory body – the HFEA, which was established then – have continued to attract public attention and criticism from some interest groups. Public consultations, or what have more recently been framed as 'public engagement' exercises, have, since the 1990s, increasingly been initiated by government to ensure both that public attitudes towards development in reproductive, genetic and other biotechnologies are sought and that greater 'public understanding' of scientific issues is generated to inform public debate. In relation to embryo research and, more recently, SC research, there has been a proliferation of public-engagement activities, and scientists we interviewed spoke of the need to engage with the public about their research.[1]

The extensive literature on the emergence of the 'new reproductive technologies' (NRTs), and especially the work of feminists, highlighted how women's lives could be transformed by access to fertility treatment that was developed following the early work on human embryos in the 1960s and 1970s. It pointed out that this work grew primarily from scientific preoccupations with animal husbandry, rather than a primary concern with women's health, and its potential to impact on women's lives became the focus of political controversy (Spallone 1989, Kent 2000). Indeed, feminist opinion was divided. Some feminists saw these technologies as enhancing women's ability to control and manage their reproductive lives, but there was opposition from others on the grounds that such technologies were pro-natalist rather than pro-choice and detrimental to women's health, and 'the new genetics' was perceived as a form of eugenics in another guise.[2] Despite the very different grounds for their opposition, paradoxically, by opposing the development of NRTs, these feminists found them selves aligned with those who opposed abortion as a social evil and who rejected the view that women should have a right to abortion.

Anti-abortionists mobilized around proposals put forward by Warnock to regulate, but permit, embryo research. The Society for the Protection of Unborn Children (SPUC) (originally formed to protest against the 1967 Abortion Act) and LIFE (established in 1970) were active in lobbying against permitting embryo research in the 1980s (Mulkay 1997) and, more recently, around amendments to the HFE Bill in 2008.[3] Other lobby groups date from the same period. In 1985, the PROGRESS Campaign for Research into Human Reproduction was launched: 'a coalition of patients, doctors, scientists and parliamentarians, PROGRESS had one aim: to make sure that human embryo research was protected by law so that IVF treatment could continue'.[4] As a pro-research organization in favour of embryo research, it was designed to organize scientists and others to promote greater awareness of the benefits of scientific research and the medical advancements that could be achieved, a message promoted by a clinician who was head of Hammersmith Hospital infertility services and who became president of PROGRESS – Professor Robert Winston. In 1990, just before the vote on the final bill, he announced that he had used genetic testing to ensure that women did not give birth to sons suffering genetic disease (Mulkay 1997).[5]

Debates about embryo research in the United Kingdom, as elsewhere, have been tied to moral concerns about the destruction of embryos – those created both as part of IVF treatment but also as a result of deliberate pregnancy termination. Therefore, although, historically, organized opponents of embryo research were organizations formed in protest at abortion, interestingly, the terms of the debate shifted. The regulatory framework for embryo research set up in 1990, creating the HFEA, separated its function and remit from the regulation of abortion.[6] The HFEA model meant that researchers wishing to use pre-implantation embryos in their research were, from 1990, legally required to seek approval from the authority and obtain a licence for their use.[7]

Interestingly, this institutional separation was reflected by a separation of abortion from the work of PROGRESS, which supported embryo and genetic research and genetic testing. Two members of PROGRESS (one with a legal background, R1; the other with a training in genetics, R2) tried to explain, in an interview in 2006, how, and why, it had no remit in relation to abortion issues:

A (R1): Well [abortion is] out of our remit because PROGRESS was set up, as the charity, as the educational trust it flowed from a lobby group which was a sort of coalition of scientists and social scientists and others and clinicians who got together to lobby for the Human Reproduction and Embryology Act, Human Fertilisation and Embryology Act in 1989–1990 to just sort of ensure that that got passed through Parliament. And the main bit of that Act is not, although it touches on abortion with the time limit, that's was not what it was about . . .

A (R2): Our sort of broader remit I suppose [is . . .] debate on genetic aspects of reproductive decision making. I guess the other area that we sort of touch on is pre-natal tests and termination for fetal abnormality.

A (R1): Because it is genetic.

A (R2): Yeah, yeah.

Q (INT1): So doesn't that draw you into the abortion debate?

Q (INT2): One of the outcomes would be termination.

Q (INT1): . . . one of the outcomes is termination and late terminations sometimes as well.

A (R1): Yes. I guess we don't really get into the sort of ethics of abortion or the ethics of having a law that allows us to do late terminations for fetal abnormality or you know where the weekly limit is set or anything like that. We're not . . . you know we're more interested in I guess, the decision if someone does have a fetus that's diagnosed with a genetic abnormality or something. The fact that that diagnosis is made and that that diagnosis would allow them to then choose.

They went on to explain that the focus on *technology* distinguished PROGRESS from other organizations that provide education around abortion. Therefore, despite promoting genetic testing and prenatal screening as technologies, and the potential for termination to follow from such tests, they were not engaged in any discussions around the ethics of abortion. Interestingly, they also had no knowledge of the use of aborted fetal tissue in scientific research or the regulatory environment relating to its use. However, PROGRESS was funded by the Department of Health to run a message board, as part of the consultation process around a later review of the 1990 HFE Act. Although, since 1992, it has been an 'educational trust', PROGRESS was also active in lobbying around the 2008 HFE Bill.[8] In effect, it became embroiled in a new round of debate that led to a new confrontation between pro-research and anti-abortionist groups as they lobbied parliamentarians who were to have a free vote on the bill.

Thus, arguments that scientific progress should be facilitated by a regulatory and legal framework won the debate in the 1980s and characterized science policy in the United Kingdom from that period. There was a growing awareness that public trust in scientists was in decline, and a belief that it could be enhanced by better regulation and greater public engagement on scientific issues (Department of Trade and Industry 2000, House of Lords Select Committee on Science and Technology 2000, Irwin 2001). The HFEA's job was both to secure public trust and to facilitate the science and delivery of fertility services. Despite criticisms and scrutiny of its role by the House of Commons Select Committee in 2005 and 2007, proposals in the draft Human Tissue and Embryos Bill (2007) to merge it with the Human Tissue Authority were strongly opposed, as we shall see.

The HFEA model, 1990–2008

As the regulatory authority set up in 1991, the HFEA licenses the use of gametes and embryos for treatment and research in the United Kingdom. It licenses and inspects clinics providing IVF and assisted-conception services. Research licence applications are publicly available, as are the minutes of the authority.[9] The authority also keeps a record of all registrations, treatments and outcomes that result from assisted-reproduction techniques. A review of the HFE Act had, during the period leading up to the parliamentary vote on the HFE Bill in 2008, engaged with a debate about the future of the HFEA and its legal basis (see Appendix 1). The role of the HFEA had been under scrutiny. In 2005, the House of Commons Science and Technology Committee reported:

> There are further concerns in that the HFEA has been campaigning, corporately, for changes in legislation. We are concerned that the HFEA has crossed the boundary from regulation to advocacy in its treatment of gamete donation. Its consultation on the Regulation of Donor-Assisted Conception published in November 2004 states that 'As the regulator of donor-assisted conception, it is not the HFEA's function to promote or encourage donation or treatment'. However, the HFEA's corporate memory is failing it. In 1998, its Consultation on the Withdrawal of Payment to Donors states that it was the HFEA's intention that 'payments to donors should be phased out in such a way as to minimize any adverse effects, particularly any reduction in the supply of sperm donors'. As a result it set up a working group to consider ways of maintaining sperm supply and increasing egg supply. More recently, in response to the Department of Health's consultation on donor anonymity, the HFEA stated that it has 'been keen to encourage a culture of altruism with respect to donation [. . .] [and] that the withdrawal of payments to donors may negatively impact on the supply of donors', which implies that donation is a 'good thing'. We have heard that membership of the HFEA has so far been reserved for proponents of assisted reproduction and embryo research. It is therefore not surprising that its individual members would wish to see greater availability of licensable activities. Nevertheless, by promoting gamete donation in its corporate publications it has acted outside its statutory remit and crossed a boundary that risks compromising public trust.
>
> (House of Commons Science and Technology
> Committee 2005: para 216)

Questions were also raised about individual members of the authority and whether they could hold views that contradicted the terms of the HFE Act. However, the committee concluded that, although the HFEA statutory role could usefully be more clearly defined in the legislation, it would not be practical to separate policy-making from its decision-making (para 218).

One critic of the HFEA and member of Comment on Reproductive Ethics (CORE)[10] described in an interview in 2006 how, in her view, as a regulator the HFEA had on occasion overreached its legal authority. She explained that, in 1994,

> CORE was founded in a moment of absolute perplexity when the HFEA issued a consultation document on the use of ovarian tissue for research or for therapy. And when I looked at the whole process of how that consultation had been conducted, I thought it was very worrying because it was, in our estimation, a very faulty exercise in consultation all round, I mean, even to the extent that the report was issued before the final deadline for consultation had been achieved. The responses were analysed in a way that wasn't very clear and the conclusions were not in my opinion particularly objective and the decisions were sort of on the basis 'well I think we could get away with this but we can't get away with that for the moment'. And I realized at that stage that the HFEA seemed to have an agenda that they were fulfilling and on that basis there were a number of other, mainly women from a variety of professional backgrounds, who were equally concerned. And that's how CORE was founded. We sat down together and said 'there's a need to build up some kind of structure to really look seriously at all the issues that are associated with assisted reproduction'. Some of the founding members have left because they went on and adopted different perspectives on the whole area but the primary principle is it is a pro-life organization. We do respect the right to life of the embryo, but contrary to what people say it's not a religious organization and it's not opposed to IVF per se, it's only opposed to destructive exploitation of the embryo. [. . .] We woke up and said 'who is the HFEA, where do they come from?'. And we produced a critique of the first reports of the HFEA at the time and also a critique of appointment to the HFEA, the composition of the HFEA. And out of that came concern about tension between a regulatory organization which is also becoming policy maker and how objective could that possibly, how acceptable could that be? So one of our current campaigns is to try and drum up some support for a non-policy making, non-regulatory but purely advisory national bioethics board that is democratically elected and respects the variety of opinions that are around. And you know the big ethical issues are discussed at that level – funded by Government but independent of Government.

CORE helped initiate and support legal challenges to the decisions of the HFEA and a campaign for a national bioethics committee that sought to highlight the statutory limitations of the HFEA in relation to policy-making on controversial ethical issues.[11] She argued that the HFEA had taken unauthorized decisions, for example in relation to 'saviour siblings':

> The Human Fertilisation and Embryology Authority has no jurisdiction over the developing embryo and that's why I find some of their rulings rather absurd. You know when they state yes you can design a baby but you can only use the cord blood. Says who? Who are they? They don't have any control or any power whatsoever on pregnancy and childbirth. It's not their remit at all. And the courts in fact would intervene immediately even on the issue of bone marrow, because you can't take bone marrow from a child without having a court order to do it. So the HFEA puts all these kind of restrictions into their licenses et cetera in areas that they have no jurisdiction. (AC2)

In fact, a legal challenge to the HFEA actions in approving pre-implantation genetic diagnosis (PGD) for a family with the inherited blood disease thalassaemia was finally rejected on appeal in 2003.[12] And, on the issue of creating animal–human hybrids, as some scientists wanted, she said:

> So we're here saying you can't get [human] eggs as we'd like them, so never mind we'll use animal eggs and create things. You know quite possibly the HFEA will say 'what a good idea'. Although in my opinion they have no jurisdiction over that, because it's not human gametes, it's got nothing to do with them. But however they might say 'what a good idea'. Professor Wilmot needs to get going so we'll let him do animal–human hybrids. (AC2)

The creation of animal–human hybrids has been highly contentious.[13] The HFEA received applications from two research teams to use animal eggs in SC research in November 2006. Following controversial discussions during 2004 and 2005 about revisions to the HFE Act,[14] and a government White Paper in December 2006 on proposals for changes to the Act, including a prohibition on the use of hybrids and chimeras (Department of Health 2006), in 2007 the HFEA carried out its own consultation exercise on hybrids and chimeras, identifying five types of embryo that contained human and animal DNA, but focusing on cytoplasmic hybrids:

> Where animal eggs are used in cell nuclear replacement, the resulting embryo is not only a clone, but is also a kind of hybrid. Known as a cytoplasmic hybrid embryo, it contains a small amount of genetic information (DNA) from the animal which is left behind when the majority of the genetic information is removed . . . However, the amount of animal DNA in the cytoplasmic hybrid embryo would be minimal – less than 1 per cent.
>
> (Human Fertilisation and Embryology Authority 2007a: para 2.8)

At the same time, in 2007, in its consideration of government proposals for the regulation of hybrid and chimera embryos, the House of Commons Science and Technology Committee reflected again on the HFEA model of regulation and its operation, reporting that there was support for its work from the international scientific community:

Professor Hui Z. Sheng of the Shanghai Jiao Tong University, School of Medicine, China, also supported the UK regulatory environment in this area of research. Professor Sheng told us that the UK is 'currently a world leader not only in embryological research and cloning, but also in policy making in this field'. She explained that the UK Government 'has established an image to be able to balance scientific development and ethical issues with confidence and vision' and that 'its regulatory policy in embryological research has provided a permissive but strictly regulated environment'. Moreover, Professor Sheng believed that 'UK policy has positively influenced the policy making in other countries, including China, Japan, USA *et al.*'.

(House of Commons Science and Technology Committee 2007: para 18)

The committee reiterated the view that a parliamentary standing committee on bioethics should be established to monitor and oversee the work of the regulatory authority and consider ethical issues relating to innovation in science. It also supported the creation of hybrid and chimera embryos for research purposes. With regard to the regulatory function and remit of the HFEA, it was generally recognized that the 1990 HFE Act did not (and could not) explicitly foresee the technical possibility of creating hybrid or chimera embryos, and so the legal framework needed to be reviewed. The HFEA decision that it did have responsibility with regard to such entitities and could approve licences for such research activity turned on the legal (and philosophical) questions of whether or not such entitites were 'human', and it was supported in its decision by the committee: 'This decision that cytoplasmic hybrid embryos can be classed as "human embryos" would bring them within the jurisdiction of the HFE Act' (House of Commons Science and Technology Committee 2007: para 68).

In January 2008, the HFEA granted licences to two research teams, one in King's College London, led by Dr Stephen Minger, the other at Newcastle, led by Professor Alison Murdoch. The decision was especially controversial because it was taken in advance of parliamentary review of the law,[15] but the HFEA considered that it had the legal authority to issue the licences, which would be subject to any subsequent changes in the law.[16]

As a result of a legal decision, the HFEA believes that it probably has the power to license the creation of cytoplasmic hybrid embryos, although the HFEA's jurisdiction over such embryos could be tested in the courts.

It is a criminal offence to create any embryo which falls within the HFEA's jurisdiction unless the person creating it has an HFEA licence to do so.

(para 3.4)

Therefore, the HFEA took the view that such research could be allowed under strictly regulated circumstances and granted the licences (Human Fertilisation and Embryology Authority 2007a, 2007b). Although, in the 2006 White Paper, the government had proposed a ban on the creation of all animal–human hybrids, a U-turn was noted when the draft Human Tissue and Embryos Bill was published in 2007 (Batty 2007).

Scientists too were active in lobbying government around the bill between 2006 and 2008, and, as Stephen Minger put it at a meeting in 2009, 'a new consensus between science and government' emerged.[17] Although previously critical of the HFEA, he and Lyle Armstrong welcomed its decision to issue the licences for their work using animal eggs in 2007 and, by his own account, they worked hard to 'get involved in the debate and to help Parliament understand the science'.[18]

In this draft bill, prepared by the Department of Health in May 2007 and considered by a joint parliamentary committee, the future of the HFEA was tied up with a proposal put forward to merge it with the HTA to form a new regulatory body, RATE (Department of Health 2007b, House of Lords and House of Commons 2007). A strong lobby emerged, including the Royal College of Obstetricians and Gynaecologists, arguing that the work of the HFEA was to be commended and should continue in the same form, and not be diluted or dissipated by the drawing together of the two agencies. Embryos were, it was argued, still to be regarded as a special case, as distinctive from other bodily tissues because of their reproductive potential and therefore deserving of special and separate protection.[19] Proposals to merge the two authorities were dropped, and the revised Human Fertilisation and Embryology Bill reflected the changed focus of the proposals – that was, to retain the HFEA model but to provide for revised and updated legislation on assisted reproduction, and for changes to the regulation and licensing of embryo use in research and therapy. Amendments to abortion law were also tabled during the passage of the bill, but subsequently rejected. The difficulties of keeping discussion and review of embryo research separate from abortion were seen as politically sensitive and hampering revision of the HFE Act, according to a senior clinician, former member of the HFEA authority and the Scientific Advisory Committee of the Royal College of Obstetricians and Gynaecologists, who I interviewed in 2006. Recognizing the need for a review of the Polkinghorne guidelines on fetal tissue research, he explained:

I can't see how they're going to revise the Human Fertilisation and Embryology Act without touching the Abortion Act because it's got to come out of the current Act. Remember that the current Abortion Act sits inside the Human Fertilisation and Embryology Act. [. . .] There are a whole lot of clauses that are sitting as modified by the HFEA 1990 that are sitting in that Act. So if the Act had ever come back to Parliament, for genuine changes to the way that eggs and embryos were to be used in patient's interest, it would have brought up that entire Abortion Act again. That's why it never came back to the House. [. . .] The Parliamentary activity to do with gametes, eggs, stem cells, everything else is completely stultified because of the politics around abortion. That's why it's kept quiet since 1990. And if the Human Fertilisation and Embryology Act came back, you would have another discussion about abortion. (CL1)

As subsequent events revealed, abortion politics did indeed shape the debate about the passage of the 2008 HFE Bill, but, as no amendments were accepted, the separate protection afforded to the pre-implantation embryo was secured. After the vote in Parliament on this bill, a well-known ethicist in the United Kingdom, Professor Julian Savalescu, posted this on his website, claiming the moral high ground:

The UK now stands as one of the most advanced, if not the most advanced, countries in the world in terms of regulation of reproduction and embryo research. Its parliament is to be congratulated. Most countries with a strong influence of religion over ethical issues have failed to make these advances. Some even ban embryonic stem cell research. Initially, the government produced a white paper suggesting banning the creation of cybrids, but after a response led by Evan Harris and proper debate, this position changed. This demonstrates how a secular ethical approach can successfully influence law and public policy for the better. What worked in this process was a range of individuals and groups making public arguments in favour of progressive reform. The UK now not only leads scientific progress in these important areas, but ethical reflection and policy formation.[20]

Although it has not been my intention here to detail the legal or moral issues at stake in the passage of the original 1990 HFE Act, nor the amendments to it in 2008, what I have sought to sketch out is the highly contentious policy context and some of the political dynamics that have shaped the regulatory framework. Importantly, at the heart of the UK approach is a view that embryo research is expected to deliver therapeutic benefits – first, to aid those receiving IVF and assisted-conception services; second, to screen for and prevent genetic diseases in families; and, third, to help a wider community to whom SC therapies might be of benefit. It is the promise of therapeutic benefits that has outweighed the moral concerns of those who object to the destruction of the embryo for research, 'therapeutic cloning' and the mixing of human and animal cells by somatic-cell nuclear transfer. What we see in Britain is a very particular national strategy for effecting the 'normalization' of embryo research, or 'for making human biotechnology seem mundane and governable in the face of moral uncertainty and conflict' (Jasanoff 2005:147).[21] Tactically, the government proceeded with some caution in its review of the HFE Act from 2004, and, by preparing a draft Human Tissue and Embryo Bill in early 2007 and releasing it for comment in advance, it created the space for further dissension, debate and generation of consensus before even bringing the bill formally to Parliament. Hence, the perhaps rather unexpectedly controversial proposal to merge the HFEA with the HTA was dropped, and, in the process, the legitimacy and support for the work of the HFEA were, I suggest, strengthened. The changing government position on the creation of hybrid embryos between December 2006 and May 2008 was like the earlier discussions in the embryo-research debate, in so far as it was influenced by the voices of leading scientists and the pro-research lobby. In order to win the argument that the creation of admixed human embryos should be allowed, it was crucial to maintain trust and confidence in the ability of the HFEA to regulate the practice. In so far as the proposals for RATE undermined that confidence, they potentially weakened the pro-research view that progress in science was, and could continue to be, strictly regulated by the HFEA. In any case, it was suggested that the authority and effectiveness of the HTA, with which it was supposedly going to merge, were untested and unproven (Kent 2009).

A national scandal: from professional self-regulation to a new regulatory order

Thus, in the United Kingdom, abortion politics framed much of the initial debate in the 1980s on embryo research, which played out as a conflict between the pro-research lobby and anti-abortion groups. In the HFEA model, IVF, assisted conception and embryo research were intertwined, but were separate from the regulation of other body parts – organs, tissues and cells. Moreover, interestingly, the regulation of fetal-tissue use in research fell outwith the HFEA, except, it seemed, in the debate about the use of germ cells (ova) from aborted fetuses in 1994 (Human Fertilisation and Embryology Authority 1994). Then it was suggested that, as there was a shortage of ova available for assisted reproduction, it might be possible to use the ova and ovarian tissue from aborted fetuses or female corpses. Even though the clinical (therapeutic) potential of using either was not certain, the HFEA sought to develop a policy position on the issue and, following a heated consultation, decided that no licences for therapeutic use of ovarian tissue from aborted fetuses would be granted, and that the issue of whether or not cadaveric eggs might be used would be kept under review. However, as the techniques for the use of these tissues were not yet developed, even if they had decided to grant licences, they could not have been used (Mulkay 1997:147). These materials could be used for research purposes with an HFEA licence. Indeed, the use of fetal tissue in research was already permitted, and there was guidance issued in 1989 about its use (Polkinghorne 1989), but no licences were ever issued for the use of ovarian tissue or ova from fetuses or female cadavers.[22]

As a regulatory object, the aborted fetus was ambiguously sited between two spheres of policy-making in Britain: although the HFEA had no formal role in relation to the regulation of abortion, because of its remit with regard to human gametes, the use of ovarian tissue/ova came under the purview of the HFEA in 1994. There were, however, already guidelines about the use of fetal tissue in research that had been drawn up in 1989, when news of trials using fetal-tissue transplants for PD came to public attention. Known as the Polkinghorne guidelines (Polkinghorne 1989), these set out guidance on the process of seeking consent and collecting an aborted fetus. The fetus was regarded as a cadaver, and women were thought to be in need of protection from researchers and from themselves. It was feared that a woman might be persuaded to have an abortion in order to donate the fetus, and so the guidance said that consent was to be obtained after the decision had been made to terminate the pregnancy, and the person asking for consent to collect the tissue should not be the person who would benefit from its use (Pfeffer and Kent 2006, 2007).

In the context of SC science, the very different kinds of protection afforded the *in vitro* embryo and the *ex vivo* aborted fetus come into view in particular ways. Explaining this in terms of the vitality and viability of the tissue and cells, while recalling the discussion in 1994 about the use of fetal germ cells, one SC scientist interviewed in 2006 commented:

A: Yeah I remember the contrasting status. I mean it is odd. You know I've always thought it's, society's odd. You know we can do experiments on embryos, research up to fourteen days and then we're terminating stuff that we could do anything on after . . . and of course it's not . . . one is considered viable, so you've got to destroy it before fourteen days because it still has the capacity for life while the other, where you're taking termination tissue, there's no capacity for life. [. . .] A pre-implantation blastocyst is viable and if it could be transplanted into a woman, then it might have the capacity for life. But it's definitely viable, it's alive. Tissue from termination, the cells may be viable but the organism has lost its viability because of the fact that you can't put it back together; you couldn't put that back and get . . . So the cells are viable. It's like after death the tooth bud still grows or fingernails still grow. There's still some viability of cells after death.

Q: Hair still grows, that's very weird.

A: And hair still grows. So there's still viability of certain cells. But as far as the organism goes, it's dead. (SS7)

In these terms, once a pregnancy has been terminated, the aborted fetus is a dead organism that contains some viable cells. Another clinician, who worked closely with SC scientists at what has been called the 'IVF–SC interface' (where IVF clinics supply embryos to SC scientists) (Franklin 2006a), remarked on the different regulatory requirements and the need for scientific investigation of all potential sources of SCs:

A: Yes. Because I don't see how you can take embryos *in vitro* to 14 days completely separately. Make one set of rules and give this extraordinary protection and give a totally different level of protection to the fetus *in vivo*. And then in the midst of all this you've got the issue of stem cells: should we be using embryonic stem cells? I don't know. Should we use fetal stem cells? I don't know. Should we use EG cells? I don't know. Should we use umbilical cells? I don't know. The answer is you've got to go down each of these avenues and try and find out what you can find out about those kinds of cells, they are all slightly different. And one of them is going to turn up probably the trumps way of going ahead. If you had a way of doing things, ideally most people would rather use umbilical stem cells because they are, for want of a better word, free, they are easily obtainable with a certain amount of clever and appropriate working, you can get them, they're varied et cetera, et cetera, they're non-controversial other than the bits we've been speaking about legally, but that can be worked around. EG cells, John Gearhart's stuff, which you were talking about with the ovarian tissue. That was his way out of dealing with his climate in the States. He's also a very sensible ethical guy. And then the adult stem cells, the bits that are sitting around in bone marrow and so on, I'm not sure that's really going to go, but who knows? And then there's the embryonic stem cells which have huge potential as we understand it. We understand what an embryonic cell can do. It makes sense. But that's also to do with a lot of science, to do with differentiation. It's not just about stem cell therapy, it's not just about regenerative medicine, it's understanding the very basis, and one of the most important, challenging things is understanding reversal of differentiation. Because if every one of your skin cells is a potential sperm, a potential egg . . . (CL1)

Here, the legality of terminating a pregnancy and using a fetus that is much older than 14 days in research is seen as contradicting the high level of protection afforded the *in vitro* pre-implantation embryo. The ethical issues were seen as influencing the direction of scientific efforts, and, in the United States in the 1990s, according to him, the decision to use the gonads and germ cells from aborted fetuses was a 'way out' of the controversy around using pre-implantation embryos (Shamblott *et al.* 1998).

In the United Kingdom, since 2004, a new regulatory regime has emerged to regulate the removal, storage, use and disposal of human bodies, organs and tissue, from the living and deceased. There was a shift from what could be seen as a form of professional self-regulation to a more centrally managed approach. Change followed along twin tracks – the first was a response to what was regarded as a national scandal at the Alder Hey hospital in Liverpool; the second flowed from national and international policy concerns about the quality of tissues and cells available for transplantation.

Using human tissues in research

In the United Kingdom, 2004 was a highly significant year for the regulation of human tissue, because it marked the passage of the Human Tissue Act, which followed the inquiry into practices at Alder Hey and the review of organ retention and storage of tissues for diagnostic and research purposes. The Act set up the HTA, with responsibility for the regulation and oversight of human tissues. Until the Alder Hey inquiry, it had been common practice for pathologists and surgeons to retain tissues taken from the living following surgery, or from the deceased at autopsy. However, when the now infamous Professor van Velzen came to public attention, and the collection of children's body parts and fetal tissue at Alder Hey hospital in Liverpool became public knowledge, there was an outcry. Professor van Velzen was found negligent in a number of respects. First, consent to retain the tissue had not been properly obtained. Second, the specimens were poorly stored and organized, and, third, the material wasn't being used. In addition to the legality of the pathologist's activities being challenged, failure to carry out research or to produce knowledge based on investigation of the specimens was considered a serious form of misconduct in this case, and he, the hospital and the university were considered culpable. One of the fetal collections – at the Institute of Child Health – belonged to the university, the other to the hospital, but limited research had been carried out on either (Redfern Report 2001).

According to this obstetrician, speaking in 2006, what had occurred was a shift in public attitudes and expectations, together with a movement away from more paternalistic approaches to clinical care to greater involvement by families and individuals in their care and decisions relating to it:

A: . . . because I believe that what happened there, besides that one crazy man, was not an obfuscation exercise but a change in general public perception about keeping tissues. If you look at practice today, would you expect to be involved and your partner or your children involved in your heart surgery? Would you like them to be present in the theatre?

Q: No.

A: No. But you might expect that for a delivery, you might expect it for treatment that is not necessarily theatre but treatment taking place on your child. And the vanguard has been obstetrics where women have taken a greater role in their own effectively normal process. The idea that it is medicalized, we've got past that, I hope.

And, drawing comparisons between changes in obstetric care and wider change:

> It's a changing mood that I would say . . . I'm going to put fifty years on that. I mean I've been at this for thirty years and it was before that. The same thing I think has happened with tissues. I started my career working in an anatomy department okay. We always had tissue in bottles, babies in bottles, interesting specimens in bottles. You did that, it was always part of what you did and you continued doing that. And suddenly the shift happened just like the shift happens in where you take part in your delivery. The shift happened without the people doing it either being aware or conscious of it. So they didn't do anything wrong, they didn't do anything different. But our view of it changed. We changed. They say, 'how dare they do this! How dare they take pieces of tissue without asking permission?' We'd always done it, done it like that, it had always been in that way where it was viewed of medical or scientific interest, teaching interest, and it was presumed that the person who had donated or taken it from really didn't care at all about it. (CL1)

Thus, presumed consent had become no longer acceptable, and consent was a key concept in the new 2004 Act. A Department of Health official put it like this:

> So I don't think it's necessarily a unique thing that fetuses or fetal material was kept, I think what became clear was that that was a position that just could not continue, that the fact that it was being kept without people knowing it, being aware of it and being quite sure of its provenance or what it was there to do is what was the problem, so what we needed to do was to bring this inside the regulated environment so people could say, 'I'm keeping this for this reason with this consent and this is what I'm going to use it for and you can come and have a look'. (PM4)

Following Alder Hey, a review of national human-tissue storage and research practices was carried out.[23] Policy in relation to donation, procurement, storage and disposal of tissues and organs was developed, culminating in the 2004 Human Tissue Act and changes in how research tissue is regulated (Mason and Laurie 2001, Medical Research Council Working Group 2001, Price 2005, McHale *et al.* 2007, Kent and ter Meulen 2011).

Tissue banking: therapeutic use of human tissue and cells

The second trajectory of change related to the clinical or therapeutic use of human tissues and cells (i.e. excluding gametes). In 2001, the Department of Health introduced a voluntary code of practice for tissue bankers that anticipated both national legislation and changes at European level

(Department of Health 2001). This Code Of Practice sought to introduce standards and, for the first time, required those storing human tissues for therapeutic use to seek a licence. It represented a significant move in that professional self-regulation was no longer considered a sufficient basis for ensuring the quality of tissues used in therapies. A Department of Health official explained the background to the code in this interview in 2003:

A: Well the first thing is that the whole thing, the history of it started with a recommendation by the Council of Europe in 1994 that Tissue Banking ought to be regulated, and that member states should look to their tissue banking to see whether it was being properly regulated or not. As a result of that the Department in 1995 set up a review of Tissue Banking in the UK run by Mike Prophet, and – I can't remember his other name now – and they came to the conclusion that it was a very mixed bag. There were some areas that were very well run and extremely professional, there were a lot of other questionable activities going on, particularly in individual hospitals where clinicians were just putting some materials, bits of bone, skin and things into fridges and then when they felt like using it they'd pull it out again and put it in without very much quality control at all. And so for a long time there was a lot of general discussion as to whether well isn't this something the profession should be doing and there was a British Association of Tissue Banks which developed in combination with the European Association of Tissue Banks, guidance on the quality and safety and sort of professional standards. And a number of tissue banks were being run by the National Blood Authority – National Blood Service – and the National Blood Service had for blood and blood products a thing called the Red Book, which is a quality manual.

Q: Which has been in existence since when?

A: For a long time, but again based on an initial recommendation by the Council of Europe which has for many years produced a guide on the quality and safety of blood and blood derivatives. And that does deal with all the quality issues and the standard operating procedures and all the minutiae of a proper quality control system in order to minimize the risk of transmission of disease from donors to recipients but also to ensure that the blood products, the blood derivatives are as effective as possible. So that's been an ongoing standard produced by the Council of Europe for it must be ten years now and the Red Book is consistent with that. So they've applied a section of the Red Book for tissue banking. So there was a great deal of professional development over the next three years of quality guidelines and standards and as usual the keen and the front runners, the people who were running tissue banks and were members of the British Association of Tissue Banking were developing much better standards but it was still very patchy. So in about 1998 it was felt by the Government that perhaps although the leaders in the field were doing an extremely good job, that there were laggards. So they set up a Working Group to develop the Code of Practice on Tissue Banking; and that Working Group finally reported and the Code of Practice was issued in February 2001. (R3)

The group was chaired and led by Dr Susanne Ludgate, Clinical Director of the then Medical Devices Agency (MDA) (see below). Previous standards were brought together in the code, and, in 2001, a voluntary accreditation system was set up to enable tissue banks to register voluntarily (there was, at that time, no legal basis for a statutory registration process). Tissue banks were to have quality systems in place by April 2003 and would be subject to audit and inspection, as the same official explained:

A: You need to keep validating your standard operating procedures to make sure that you are delivering safe effective [inaudible] routine testing procedures, for instance for microbiology and safety. So you will either – for example you may have a continuous process of drop plates in your clean facility that every morning you put down new plates for bacterial monitoring and you put – you routinely culture them and you see that nothing's developing, now that for instance is a high quality micro-biological safety standard operating procedure, you put plates in every day and you culture what you get to make sure you're not getting pathogens developing. Things like checking safety cabinets and things, checking fridge temperatures and so forth, but also actually checking that the stuff, that having processed it, actually works and is still viable, or non-viable depending on what you need to do. So a quality system puts in place – and it doesn't just cover the actual processing, it covers things like packaging, labelling and so on and so forth, the whole process. (R3)

In the process, the concept of a 'tissue bank' itself came under scrutiny and was understood as far more wide ranging than was envisaged by some clinicians. Notably, the widespread practice of keeping tissues, such as skin, in a fridge on the ward of a plastic-surgery unit came within the code, as they explained:

Well, quite a lot of banks, quite a lot of establishments didn't recognize they came under the Code of Practice, so for instance bone marrow transplanters didn't think they were covered because everybody sort of thinks of bone marrow – of tissue banking – as the traditional tissue banks, heart valve, bone, skin, tendons, corneas – you tend not to think about the storage of blood cells, cord banking, things like that as being part of this. And even more – people like orthopaedic surgeons who feel they just want to store some bone, and plastic surgeons who treat patients who will take skin grafts and graft the patient, but they'll keep a bit of skin wrapped up in the fridge in case it fails so they can patch it afterwards – and all those people never dreamt for a moment that they would ever come under the Code of Practice because they didn't see themselves as tissue bankers. (R3)

A tissue bank in this context was anyone storing, processing or distributing tissues for transplantation – for reimplantation. The code didn't cover tissue used for laboratory research, but clinical research, where tissues were being transplanted into patients, was included. Cells and cell lines for gene therapy or commercially manufactured products using human tissue, cells or cell lines were also excluded.

Q: So can you give me an example of a situation where that kind of clinical research would be covered by the terms of the . . .

A: Well there's an area, it's sort of on the fringes of research plus tissue culture plus – or rather cell culture, tissue culture, that's been going on for some years now which is taking keratinocytes – skin cells – again from burns patients – culturing them with 3 2 3 or 3T3 mouse feeder cells to grow up sheets of keratinocytes. And you then – for burns patients you then transplant those sheets back on instead of a skin graft. So for severely – you're growing up that skin and re-implanting it. Now that hits all sorts of problems. It's still in many cases a research technique because it's not a routine procedure by any means, it's done in a lot of cell culture laboratories. They're extending the applications to eyes and other segments, they're looking at trying to culture a lot of other skin cells so most of that's going on in clinical research laboratories with some therapeutic products being produced on an ad hoc when/if necessary basis. Now that particular activity, they never thought of themselves as tissue bankers at all because they're researchers. The second thing is they didn't recognize that if we change, if we come into line with the United States and change the definition of xenotransplantation those cultured cells have been in contact with mouse cells so it's also a xenotransplant and they didn't recognize that because this is something they've been doing for ten or fifteen years.

Q: OK so what's the implications then of that?

A: Well the implication of that for that particular area clinical/research plus – some of its research, some of its actual treatment – therapeutic application – is that those laboratories should all come under this guide and they should – well if we change the definition of xenotransplantation to come in line with that used in the States and now being recommended to the rest of Europe then that will be a xenotransplant procedure and they'll also have to apply to UKXIRA for accreditation. So there'll be double regulation on them, where at the moment they see themselves as totally unregulated. (R3)

These kinds of activity, then, were at that time unregulated, but would from 2002 fall under the voluntary code. Whether such products should be regarded as a type of 'xenotransplant' was hugely contentious and was the subject of a separate report to the UK Interim Xenotransplantation Authority (Shin and Griffin 2002, Brown *et al.* 2006a).[24] They were at the boundaries of existing regulation and raised concerns about cross-species contamination.

Hence, following the adoption of the EU TCD described in Chapter 3, when it opened for business, the HTA had a wide range of responsibilities relating to the regulation of human tissues. These included oversight of those both collecting and using human tissues in research and therapies. Although wide, its remit did not extend to the collection and use of human embryos or gametes, which, as we have seen, were regulated separately by the HFEA. The regulation of human tissues and cells for research and therapy in the United Kingdom was rooted in the institutional and political histories of these two quite distinct regulatory bodies. The HFEA, as we saw, provided a 'strict' but permissive approach to embryo research and, more recently, the creation of human–animal hybrid embryos. In contrast, the HTA was established as a response to national concerns about the donation, use and disposal of body parts from the living and the dead. Its task was to develop a more centralized regulatory system to control a much wider range of activities, both under the terms of the Human Tissue Act and as a competent authority under the EU TCD.

However, the HTA welcomed proposals to merge with the HFEA, saying in a statement responding to the White Paper in 2006:

> Bringing all matters concerning human tissue, gametes and embryos under a single framework will ensure consistency of approach in these closely related areas. It also makes sense in terms of European legislation – the HTA and HFEA are both Competent Authorities under the EU Tissue and Cells Directive (EUTCD).[25]

Shirley Harrison, Interim Chair of the HTA, said:

> We have worked to establish the HTA as a regulator that uses a modern and creative approach. We are fully supportive of RATE and want to see the new Authority combine the best of both the HTA and the HFEA to become a model of better regulation.
>
> (ibid.)

She was appointed Chair of both the HFEA and HTA, with a view to facilitating the merger when the former HFEA Chair, Suzi Leather, stood down. However, it never went ahead, and one of the criticisms of the proposals and reasons for abandoning the merger was that resources were seen as already 'thinly spread' in the HTA, and there were concerns that a new authority would find it difficult to cope with an even wider remit. In response, the HTA acknowledges that its first year of operation has been a challenge: all key elements of delivery have been achieved, but we recognize we have had to compromise in certain areas because of the complexity and size of the workload and our widening remit.[26]

In addition to 'building confidence' with its stakeholders, the HTA portrayed itself as an emerging authority that was still in the process of developing appropriate systems to implement its regulatory functions, from 2006 onwards. In the wake of the government decision to drop the proposals to merge the two authorities, Britain's implementation of the European TCD became the responsibility of these *two*, separate, competent authorities. This was somewhat anomalous, as, within the terms of the directive itself, the donation, procurement, processing and storage of gametes, embryos, cadaver tissues and tissues and cells from the living, cell lines and stem cells were all covered by the legislation. More commonly, elsewhere in Europe, implementation of the directive falls to a single competent authority.[27]

In addition, under the terms of the new RATMP, in the United Kingdom, the Medicines and Healthcare Products Regulatory Agency, itself a 'hybrid institution' (Brown *et al.* 2006a), assumed responsibilities where tissues and cells were to be used for therapy and were subject to an 'industrialized' process (see Chapter 3). For scientists in the laboratory, the regulatory hurdles were especially complex and confusing. This scientist described the difficulties of obtaining accreditation for a new laboratory built to GMP standards:

A: This is a minefield, oh my goodness! We've been thrown from every regulatory authority. I mean it's just crazy.

Q: Tell us about it.

A: Well it's outside the HFEA 'cause they don't want to know about it. It should be the responsibility of the MHRA but they've been . . . I don't know what they've been doing, they've just been holding . . . I mean we could have been accredited a year ago. But they've been saying one thing and then another. And then it was the HTA. But as far as I know, embryonic stem cells are outside the HTA remit. This is my interpretation but I'd love to hear your interpretation of this.

Q: So did you apply to MHRA for accreditation?

A: We have done yes. We've gone to them several times and they've ummed and erred and they've really just not responded to us. So then eventually we have had a response saying they will do it under a voluntary code and they will now accredit us. (SS7)

And

A: Just in terms of embryo research. So if I take a four cell embryo and grow it up to the pre-implantation stage, that is their [HFEA] responsibility under the Act. So I have a licence from the HFEA which allows me to do that and I am the person responsible to do that. The moment it's destroyed, that's where they end really. They're not interested what happens after that because that's their point of . . . the embryo's destroyed. What happens after that is not their responsibility. So then you have an embryonic stem cell line. You could argue that that line is no different to tumour cell lines produced from any other tissue. [. . .] So I do come under the HFEA but only up to that [point]. For the GMP, that is MHRA and possibly HTA. But the HTA aren't interested at the moment. (SS7)

This interview, which took place in 2006 before the new EU RATMP had been finalized or came into force, revealed the ways in which the permeable and unstable boundaries between the HFEA, HTA and MHRA were being negotiated. It was a frequent complaint at scientific meetings between 2006 and 2008 that none of these authorities could, or would, clarify its responsibilities, particularly in relation to the accreditation of SC laboratories. In two later position statements, the authorities sought to identify and fix the boundaries between them. The HFEA had responsibilities for the use of embryos but, once the SC lines had been derived, had no remit. 'During the cell line derivation process the embryo is dissociated and it is at this processing stage that the HTA regulatory remit begins and the HFEA's regulatory remit ceases.' And, 'once Master Cell Banks have been created with a reasonable expectation of clinical utility in a medicinal product, they fall within the remit of the MHRA.'[28] Later, the regulatory authorities, the Department of Health and the MRC produced a map of the regulatory pathways and a series of 'toolkits' to assist researchers to navigate the regulatory framework.[29]

Regulating hybridity and boundary objects

As we saw in the previous chapter, ethical controversy and international debate about the 'commodification' of the human body played a key role in shaping the regulation for new tissue- and cell-based therapies. The implications of defining human tissues as 'products' of some kind were, at least for some member states and stakeholders, especially problematic, because, symbolically, it first underplayed attributions of personhood and humanness and, second, potentially unleashed the marketization and commercialization of human body parts. What emerged was a parallel but closely intertwined regulatory system, where the EUTCD focused primarily on requirements for donation, procurement, processing, storage and distribution and sought to uphold the ethical principles of voluntarism, not-for-profit procurement and gifting of tissues, informed consent and do no harm to recipients of transplanted tissues and cells. Meanwhile, the EU RATMP was principally concerned with bringing these new therapies to market; product safety; and promoting harmonization of standards in order to facilitate the commercial exploitation, diffusion and expansion of the use of these technologies within health-care systems. It was also consistent with the aims of supporting EU science and securing a European stake in the global market for emergent biotechnologies. Dispute had focused on how these technologies, which combined biomaterials (for example, synthetic scaffolds) and human tissues, and cell therapies could be categorized and regulated. Were they 'device'-like and should be regulated as devices, or were they more like drugs and could be accommodated within pharmaceutical regulation? Or, was a third pillar of regulation specific to tissue-based technologies needed? These boundary disputes were played out at the national level too, and, in the United Kingdom, a strong view of the advantages of a lighter-touch, device-like approach was favoured in early discussions. This in part related to the way in which the issues were taken up by the then MDA, which represented the UK position in the international policy arena. The MDA was set up in 1994 as the competent authority responsible for implementing the EU MDD, but it merged with the Medicines Control Agency in 2003 to form the MHRA (Kent and Faulkner 2002, Faulkner *et al.* 2003). The merger was itself a response to the problems of borderline issues relating to health-care products.

From 2000 onwards, within the United Kingdom, regulators noted the 'regulatory gap' between drugs and devices and, in light of the exclusion of products containing human tissues from the device regulation, argued that there was a need for a third approach (Cox 2000). In the absence of agreed European regulation, until 2008, member states regulated the new tissue-engineered products differently, so that the same product could be regarded as a medicinal product in one country and as a device in another, or it could be completely unregulated (Faulkner *et al.* 2003, Kent *et al.* 2006b). In the United Kingdom, in 2002, a Voluntary Code of Practice for the use of what it called 'human-derived therapeutic products' was drawn up by the MDA, in consultation primarily with industry representatives.[30]

> Human-derived Therapeutic Products are products that use material of human origin, and which are used for therapeutic purposes. These may include products such as bioengineered skin systems, cartilage repair systems, and novel bone substitutes that may use scaffold matrices with autologous or allogeneic human cells.
>
> (Medical Devices Agency 2002:3)

Although it had no statutory force, it was designed to promote safety and confidence in the new tissue- and cell-based products and facilitate their distribution within the United Kingdom, together with the UK Voluntary Code for tissue banks (described above) and guidance on microbiological safety (Department of Health 2000, 2001). In this interview, in 2003, with a senior member of the MDA and the newly formed MHRA, he explained:

A: In some member states, some national rules had been brought forward – probably could be contested and certainly you can't, under the European rules, bring forward your own legislation if (that is currently), there is a proposal for the European Commission. Now if you go back to around about late 2000, well in October 2000 the Scientific Steering Committee produced their report under David Williams and what that report says in effect is there should be a third way that these products will, yes be regulated neither as medicines nor as medical devices but by a third way. It sets out some principles of that.

Q1:   This is the SCMPMD?

A: Yes, the Scientific Committee under DG Sanco. And I think that found quite broad favour, certainly in the UK, that this was the right way to go – third pillar legislation.

Q2:   Is that an influential committee?

A: Yes it is. I mean obviously it's a very well respected committee of the Commission, has a lot of influence over there and was well received at the time. We highlighted, the UK ministers highlighted in some speeches around there that there was a problem, there was a need for regulation because we've got products on the market which are not being controlled. We therefore decided in the UK to produce the guideline which you've seen, which was an attempt to bring some informal control on the area, in effect setting out best practice, what manufacturers should do in the absence of statutory controls. That was well received both nationally and I have to say, internationally and has been taken forward at European level. So a lot of people are following that and certainly we've discussed that through the international harmonisation group. (R2)

In a sense, it was an attempt to enable the industry to regulate itself in the British context and to set out a position from which to negotiate the terms of the European regulation.[31] Consistent with the device regulation (Directive 93/42/EEC) and close relationships between the device regulator (MDA) and industry, it put the responsibility on the manufacturer (producer) to demonstrate the safety of its product and the efficacy of quality-assurance and risk-assessment procedures, which could be subject to audit.

Q1:   And in terms of how that's operating, can you explain, obviously it's voluntary for the industry, so what does that mean really in practice?

A: Well what it means is that it sits there. I mean, do we police it? No we don't. Could we police it? No we can't, because of course given these products aren't regulated, we don't [. . .] look at the products when they come forward to the market on there. So it is purely a voluntary guide and I understand that it is certainly being advised because it brings together best practices and very simply companies want to follow best practice, they don't want to be sued, they don't want to have safety failures. So it's as simple as that. But do we monitor it? No we can't because these products aren't regulated. Even if they were, and of course that would be perhaps a Notified Body[32] responsibility so we would be at arms length, but we're not even there. (R2140503)

Public-health bodies, consumers or patient groups, clinicians and academic scientists were only weakly represented in these early discussions about the need to ensure the safety of the new technologies. The drive towards commercialization and promoting trade was a priority at the national level (Kent *et al.* 2006b). There was, however, considerable uncertainty about both the kinds of evidence that should be required for marketing approval of such products, and the stabilization of concepts within the field. The term 'human-derived therapeutic product' was itself conceived of in the development of the code, in the absence of any agreed scientific or regulatory terms to encompass the diversity of products expected to emerge. As with the later international debate, demarcation issues were difficult to resolve, and ethical sensitivities relating to 'human' products were of particular concern, although regarded as outside the remit of the national regulator (Faulkner *et al.* 2006).

In research that was focused on skin, bone and cartilage tissues, the regulator, when asked about the marketing of these in the United Kingdom and the safety issues, responded:

Q1: Is a consensus about the kinds of evidence and testing needed to evaluate those technologies at this point or not?
A: No, I don't think there is. In terms of effectiveness or performance in there?
Q1: Well in terms of both and also safety.
A: Well I don't think there is you see. I think the issue that would have to be looked at there is in terms of are you looking at short-term performance or long-term performance and that is a difficult area. You're looking into, what are going to be our benchmarks? Is it short term, 28 day? Or are we looking at long-term performance in there? Are we looking at the quality of the skin if it's a skin graft? What are our criteria? And then you come back to the claim you see. Is it to aid the natural healing process? Is it to produce repair to a full-thickness burn on there? Is it the quality of the skin? What are you actually looking at? In terms of safety I think for many of these products we're into another era because what I was looking at is long-term safety. And therefore, just as for many of the biotechnology products, we're not looking at our present systems, which are by and large built around spontaneous reporting: there is a problem – we report it. It's probably actually looking at proper cohort studies, proper long-term surveillance and saying are there any risks attached with this five years downstream? What might emerge? And those are the sort of issues we've

got to look at and that may take a time to come forward. Is there a risk of viruses crossing over? Is there any oncogenic expression in there? What is the safety of this? Have we changed the genetic profile and so on. Those are the sorts of issues'. (R2)

The regulator recognized the dangers of 'regulatory capture' in a comparison of the medicines and devices sectors:

Q2: So thinking of putting sets of guidelines for those sorts of types of evidence together, who are the main stakeholder groups who need to come together?

A: Well the normal way of doing this is, it differs between the two sectors. In the medicine sector it tends to be that you'll go to academics, advisory committees in the UK who would turn to the CSM for example or other specialists and we'd use that with expert panels to bring forward a proposal, and then we would put that out to the industry for comments, and to academia to have perhaps one or two rounds of consultation. That's very much the frame of determining guidelines on the medicine side. In the devices area it tends to be well let's do it jointly between the industry and regulators/academics. It has its advantages but also its disadvantages. I think in a way the medicine side is perhaps rather better because it allows comment on there but it allows people without interests to actually formulate what is right, in terms of public health and then expose that for an honest debate, people saying well yes what a wonderful standard but it's impractical it can't be delivered and so on and so forth, have you thought of. . .? Rather than the devices side rather more jointly developing them where there is a slight danger of some capture coming in there. That's a matter of process in there. But how do you do it? Well you've got to bring together regulators, academics, the commissions concerned and get a consensus about what are we looking for, what is reasonable, what is proportionate to establish the evidence base that this product works in the terms described. (R2)

So, central to discussions at MDA (and MHRA) were concerns about risk and patient safety, which were framed as technical problems relating to the classification and identification of risks and understanding of how the new tissue- and cell-based therapies acted in the bodies of recipients:

A: What I was saying there is that the new definition of an 'advanced therapeutic medicine' has changed and it refers to the product which has a physiological effect. Now I know where that comes because certain medicines act on physiology rather than on pharmacology. But if you're talking about it in terms of cells having a physiological function, then hang on, if you have a group of cells which you then call a tissue, then they work physiologically because they cover a surface. And that's where the definition could start to cause some difficulties. I can see lawyers arguing, oh well this means my product's definitely a medicine. Or then having to argue well, it doesn't mean that. And that's the difficulty. And the word metabolic doesn't help either because every viable cell metabolizes, otherwise it's non-viable. And some tissues metabolize as well, so that gets even more of a complication. And the real truth is that a lot of our legislation was written when it was very simple. You had immunological products – vaccines, metabolic, OK, well those were certain sort of food preparations given parenterally and so on, and you and pharmacological actions. We all knew what pharmacology was didn't we? But then when you get down to the minute receptor level, then is it really pharmacology and how is the cell operating? Is the cell actually working by elaborating cytokines? It gets horribly messy at that level. (R2)

It was suggested that, in the United Kingdom, there was greater transparency and openness in policy-making compared with other member states, and there was a willingness to promote greater public participation (including expert patients) and involvement of clinicians (one explanation given for setting up the new Committee in Safety of Devices in 2001). This was attributed to the recent history of public-safety issues and the decline in public trust of expert knowledge:

A: So the questioning of the expert and so on. I think it's much more advanced here. It goes back to sensitivities. I think maybe through BSE and other episodes the UK's much more questioning of experts. But it always has been. (R2)

In practice, however, there was quite limited public participation in discussions during this period.

In the international debate, the staff from MDA (and subsequently from MHRA) were directly involved in the discussions with the EC about developing new regulation and claimed a leading role in shaping the direction of those developments that were initiated by the devices sector within the Commission:

A: This is discussed at the Medical Device Expert Group, which the UK is a leading player in.
Q1: That's within the DG Enterprise?
A: The Enterprise area, yeah. This is a subject which we've discussed two or three times there. There have been some small groups established

which I've sat on some of them with other colleagues, but Didier Bouis has been leading in taking this forward. So there are groups discussing what we call non-papers. There are informal meetings, and the Commission brings together its proposals. UK ministers who have made quite a few statements on this, our guideline is very much shared with European colleagues. So we are in Europe up to our neck on this one, while the items we were discussing with the Commission obviously trying to (this coming Friday) in Brussels. So we were involved in that and probably leading the debate I suppose it would be fair to say. (R2)

In addition to the engagement of MDA/MHRA staff in discussions with the Commission, another influential figure from the United Kingdom was Professor David Williams, a scientist from the University of Liverpool and UK Centre for Tissue Engineering and chair of the Scientific Committee on Medicinal Products and Medical Devices (SCMPMD). This committee was set up as part of the advisory structure to provide scientific advice to the DG Sanco in the wake of international public-health scares around BSE and other contamination.[33] As chair, Williams was instrumental in promoting policy-making around human-tissue engineering and in preparing its Opinion on the issue in 2001 (Scientific Committee on Medicinal Products and Medical Devices 2001) and assisted in drafting the DG Enterprise proposals for a new TE directive. SCMPMD put forward the view that a third pillar of regulation was needed, and that neither existing drug nor device regulation would be able to cope appropriately with the kinds of product that contain viable human tissues and cells. According to it, reliance on conformity to the essential requirements of the MDD and the limitations of notified bodies' competence were problematic. It also considered reliance on a quality-assurance process inappropriate and drug clinical trials too costly and unsuited to customized TE products, and it was this position that the UK regulators supported. However, as we saw in the last chapter, the political dynamics between member states at the EU level led to a different outcome, and the new RATMP brought these products within a (modified) pharmaceutical frame. Crucially, a centralized procedure for market approval via the EMEA emerged, and the future role of MHRA was defined in relation to that.

The MHRA acquired new responsibilities under the terms of the EU RATMP in 2007, set out in a joint position statement with the HTA:

Manufactured products that are classified as medicinal products by the MHRA or EMEA will be regulated under the Quality & Safety Regulations only for the donation, procurement and testing of tissues and cells. The subsequent stages, including manufacture, storage and distribution, will be regulated by the MHRA. Unless exempt, ATMPs will require a marketing authorisation granted by the European Commission (the 'centralized procedure') with the EMEA co-ordinating the application and assessment procedures, and post authorisation supervision. Treatments involving human tissues or cells that are not medicinal products will continue to be regulated by the HTA under the Quality & Safety Regulations for licensable activities. This also applies to ethically (but not MHRA) approved clinical trials involving the use as grafts of human tissues and cells in patients. In cases where the regulatory status of a manufactured product derived from human tissues or cells is unclear, the MHRA should be asked to determine if the product is a medicinal product. The decision of whether a given treatment falls within the scope of the European definition of medicinal product (MP) and ATMP, as opposed to a tissue or cell graft (which would then fall under HTA remit), will eventually be determined by the EMEA working in conjunction with the MHRA (where the manufacture occurs in the UK). A Committee for Advanced Therapies (CAT) based at the EMEA will advise whether a product falls within the definition of an ATMP.[34]

Therefore, the regulation of emerging tissue- and cell-based technologies fell within the remit and scope of the MHRA, but at the boundaries with the scope of the HTA and HFEA. The boundaries between these regulatory institutions at the national level had become sites where new innovations and techniques became the locus for policy implementation and for determining how specific practices and cases can be regulated.

Another scientist explained things like this:

A: One of the problems with the HFEA in terms of how you might want to regulate this is if you want regulation over ES cell research. Okay, now I, and I think that, in itself, is an area for discussion. The HFEA has jurisdiction over pre-implantation embryos, and doesn't have any jurisdiction over ES cells.

Q (JK): No . . .

A: Because they're . . .

Q (JK): . . . or cell lines.

A: Yeah, because they're not embryos, okay. Now, given that they're not embryos, and to quote another researcher around the country 'they're only cells' and as a scientist I have to agree. Once they are ES cell, they are a cell, they're not an embryo. How much regulation do you need, okay? Why do you, why should you have more regulation, over that cell than say a fetal hepatocyte that I'm growing, or a cell from someone's liver biopsy, an adult liver biopsy that someone has got consent for taking some of those cells into culture. Why do you need more?

Q: So does that imply a kind of . . .

A: There's a potential for a double standard.

Q: . . . harmonization?

A: Yes, so I think, yes I think there might, we might see a harmonization between people like the HFEA as a regulatory body and people over-seeing human fetal tissue, or human tissue usage, okay. (SS2)

The emerging field of SC research drew attention to the very different regulatory approaches applied to different types of tissue from which SCs might be derived. As we have seen, the HTA had its origins in a very different regulatory history that focused on cadaver tissue and tissue banking and was distinct from the work of the HFE, whereas the MHRA's institutional form emerged from two (disparate) cultures of pharmaceutical and device regulation. Industry had been well organized in lobbying in Europe and enjoyed close relationships with the UK regulators (especially within the device sector), but the implementation of the new RATMP has hardly begun. Industry leaders commonly say that regulation is welcome and much needed in order to bring innovative products to market, which is why they work towards shaping the regulation. The jury is still out on whether the new structure, including a centralized approval system for products that are produced in batches or have 'a high degree of manipulation', is going to be effective. In the United Kingdom, MHRA has to approve clinical trials using allogeneic (donated cells) SCs – a phase-1 trial in stroke patients initiated by ReNeuron, whose product is derived from fetal tissue, was one of the first to be approved.[35]

Conclusion

In this chapter, I have sought to describe the national context in the United Kingdom and to explore the political and social dynamics that have shaped and structured the regulatory regime since the 1980s. New technologies that extract tissues and cells from bodies for research and for the development of treatments have been seen as challenging existing legislation and as provoking national controversy, especially in relation to the use of embryos, ova, fetal tissue and animal cells for SC research. My focus has been on the regulation of gametes, embryos and human tissues, and I haven't included the parallel development of biobanks (population genetic databases) or their governance, nor discussed gene therapy or the implementation of the Clinical Trials Regulations. However, these overlap and are also important aspects of the context of regenerative medicine. Since May 2008, amendments to the Clinical Trials Regulations have led to new arrangements for the way the UK Gene Therapy Advisory Committee (GTAC) will operate. Its role from that time was to oversee the ethical conduct of clinical research using SC lines, and in order to facilitate this most gene-therapy proposals were expected to be transferred to other research-ethics committees in the NHS.[36] The NHS Research Governance framework was developed and came into force in 2001 (Kent and ter Meulen 2011).[37]

The HFEA emerged in the 1980s in the context of a national debate about embryo research and the need to regulate fertility treatments. It has been largely supported by the pro-research community, including scientists, and the clinical professionals it regulates (for example, the Royal College of Obstetricians and Gynaecologists (RCOG)), and it emerged as a model that became a source of national pride and a means of securing public trust and confidence, in spite of criticisms regarding some of its decisions. In debate around proposed revisions to the HFE Act between 2004 and 2008, support for continuation of the HFEA model was mobilized, and it was argued that research using human–animal hybrids in particular could be best controlled within that model.

The HTA arose from different origins much more recently, with a wide remit to regulate human tissues obtained from both the living and the deceased for use in a range of different contexts, in the wake of public concern about clinical practices. Its role in relation to regenerative medicine can be seen as linked to the use of tissues and cells both in research and in the clinic. Between its establishment in 2005 and the time of writing, in 2008, it was regarded as still seeking to build relationships with a diverse range of 'stakeholders' in order to develop confidence and trust in its ability to regulate the different sectors. I suggested that it marked a shift from professional self-regulation to greater state control and oversight. However, its 'unproven' track record and lack of support from any constituency meant proposals to merge it with the HFEA were seen as threatening and undermining the reputation of the latter and so were dropped.

The MHRA could be seen as a hybrid institution that regulates across technology sectors – pharmaceuticals and devices – which were historically quite different in their regulatory approaches. Authorization and approval to market new pharmaceuticals have been characterized by long lead times, expensive clinical trials and a European centralized approval process. In contrast, medical-device regulation since the 1990s has adopted a 'new approach' to regulation, with harmonization of standards to promote a single market, and innovative devices were either self-certified by manufacturers as compliant with the regulatory requirements or approved by licensed 'notified bodies' contracted by the manufacturer. The principal functions of the MDA (and later the MHRA) were to license the notified bodies, approve clinical investigations using devices and oversee post-market surveillance. The two regulatory authorities, MCA and MDA, were merged in response to an increase in the number of combination products – those that didn't fall entirely into the regulatory category of either 'drug' or device. The first-generation products containing tissues and cells were therefore relatively unregulated until the MHRA took on formal responsibilities for them as the competent authority responsible for implementing the new EU RATMP in 2007–8. Relations between the regulator and industry have been of critical importance in shaping policy, as we saw. However, following earlier suggestions of a more user-oriented approach in relation to other implant technologies (Kent and Faulkner 2002), although the potential value of SC therapies to specific patient groups has been a focus of attention, wider participation in discussions relating to tissue- and cell-based therapies has been limited. This may be, in part, because few such therapies have yet to reach the clinic.

In sum, in each of the three areas of regulation discussed, the interests of different constituencies or stakeholder groups have been played out. In relation to the HFEA and embryo research, the interests of scientists and clinicians (the pro-research lobby) have won through; in relation to the HTA, state authority has weakened professional self-regulation, while, in relation to MHRA, industry–state–corporatist alliances have been characteristic. Just how these three areas intersect and interact in the emerging field of regenerative medicine will be important for the ways in which new therapies reach, and are distributed within, the health-care system, both in the United Kingdom and elsewhere. What I suggested in Chapter 3 was that there was a convergence between tissue-banking and industry sectors that becomes evident at the national level through the negotiations and interactions between these groups and the regulatory oversight and governance arrangements. Specifically, tissue and cell therapies became subject to the gaze of these authorities and fell within these jurisdictions according to (a) the origins of the starting material – an embryo, fetus, cadaver, living person; (b) the context of 'donation' or procurement; (c) the location and storage of the biological material; (d) the type of processing or manipulation the material undergoes; (e) the intended distribution and use of the product – whether for one recipient (autologous) or many recipients (allogeneic), within a hospital or on the medicinal-product market. A new draft 'route map' produced by the regulatory authorities was intended to guide researchers and manufacturers through what may best be described as a regulatory maze and subsequent 'toolkits'.[38] What I hope to have shown is that the background to this map is a complex political history that has shaped how new biological entities are named, categorized and regulated. The regulatory ordering of these biological entities creates new kinds of regulatory object: 'embryo', admixed embryo, fetus, human-derived therapeutic product or advanced therapy medicinal product.

More recently, in July 2010, following a government review of 'arm's length bodies', proposals to close both the HTA and HFEA emerged as part of a wider 'bonfire of quangos' (Department of Health 2010). The Academy of Medical Sciences also conducted a consultation and review of the regulation of research in the medical sciences and concluded that a new 'health research agency' should be established that could take over the research-regulatory functions of the HFEA and HTA. It was also critical of the MHRA implementation of the clinical-trials regulation (The Academy of Medical Sciences 2010). Interestingly, as the HTA pointed out in a response to this report, the separation of the research function from the other activities (which it had been suggested should transfer to the Quality Care Commission) could increase the 'burden' of regulation (Human Tissue Authority 2011). Boundaries, therefore, continue to be contested and disputed, and it is uncertain how different interests will shape the future governance arrangements, which, in any case, go beyond the formal regulatory structures and regime building (Faulkner *et al.* 2006, Faulkner 2009a).

5 'Football fields of skin': a masculinist dream?

Introduction

So far, I have explored the ways in which tissues and cells have become detached from bodies, that is, how they have been collected from cadavers and the living, and circulated or distributed as exchange objects between the clinic, mortuary, laboratory and manufacturing company. In the new tissue economies, tissues and cells potentially acquire new forms of value, becoming disconnected from the bodies of 'donors', relocated in time and space, disembodied, suspended, reanimated and reimplanted into different bodies (or the same body at a different time) (Cooper 2006b). When seen as a process of commodification, this raised ethical concerns about how such practices undermine notions of personhood and what it means to be human. These

Figure 5.1 A CompacT SelecT CellBase flexible automated cell-culture system
Source: Picture by permission of TAP Biosystems

concerns were, in turn, important in shaping international regulatory policy and the emergence of new European legislation to control these economies and facilitate the growth of 'regenerative medicine' and a market for these innovative biotechnologies. The United Kingdom has been especially keen to promote innovation in this area, and the political and institutional context described above highlights how new regulatory objects were created; how embryos and fetuses are regulated differently; the ways in which the potential for life, and 'death', are conceptualized in regulatory terms; and how (bio)ethics and regulation intersect. In Chapter 2, it was suggested that 'regenerative medicine' is seen as representing a paradigm shift, a new therapeutic approach to promote organ and tissue repair and regeneration, and implies a new way of thinking about the body. In this chapter, I want to explore in more depth what kinds of body are enrolled in regenerative medicine, how bodies are configured by these technologies, and what kinds of body are (potentially) produced. I want to develop a feminist analysis of how gender and these new tissue and cell technologies are connected. By drawing on feminist theory and science and technology studies, the ways in which gender and technology are mutually shaping and constitutive come into view. This leads to a number of important questions: first, what kinds of body are enrolled in regenerative medicine as sources of 'raw materials' or, in industry parlance, 'starting materials' – the tissues, cells or body parts that provide the biological materials for processing? Second, how are technical systems and practices for processing, manipulating, storing and production of therapeutic product/therapy gendered? Third, how are *users* of the technologies configured/gendered, and how may patterns of access, distribution and availability of treatments be understood? Fourth, what is the transformative potential of these technologies – how do they transform or re-inscribe specific social relations?

Gender, science and technology

My starting point for this discussion is an interview, in 2004, about a company called Novartis, which, in partnership with Organogenesis, marketed one of the first (and, it has since been argued, most successful) tissue-engineered skin products, Apligraf (see Chapter 1). The process of manufacturing Apligraf was described in an account of the product history up until that time:

A hundred or so neonatal foreskins were collected from a Jewish hospital and cell banks set up in the US. Fibroblasts and keratinocytes are kept in separate cell banks. Working cell lines are set up from the master cell banks for the production process. Once the production of a batch begins it cannot easily be halted, so timing of the process is critical and has to be determined by planned delivery dates. The two cell types are proliferated and then layered together with the matrix which is bovine derived. Then it is exposed to oxygen and 'lifted' to produce the skin-like surface – this enables the keratinocytes to be transformed from a gel-like substance to the qualities and appearance of skin. The product is then 'harvested' and packaged for shipping. It is an allogeneic product but has nothing in it that provokes an immuno-response. (Interview notes, M-EU6 [M-EU20])

When asked whether there were any social or ethical issues associated with these products, it was suggested that the ethnic origin of the donor of cells might be an issue. This manufacturer also said that there was a joke in the company that from one foreskin you can produce enough skin to cover a football pitch. This was a matter of some pride. Mothers of donors were screened and asked for a full medical history and about the paternity of the child. In another interview, a UK regulator said, when asked the same question:

Very early fetal cells are still very plastic and that's why they're used. A small number of cells are used to grow up fields of, literally, fields of skin. Now to that extent I think in the public mind they get a little bit perhaps worried by that. So I think it's more about perception and it's about those words tissues and engineering, and what's going on and so on. But actually if you stop and think about it what you've got is the growth of [inaudible] – so it's quite mature cells. Immature in one sense but in terms of their overall development, they're not stem cells for example. So I think it's more a perception, isn't it, that we are manipulating tissue in there. And I suppose the other fact is that at the end of the day if you take some of these skin products, this all comes from, a small piece of fetal penile skin. That's where it comes from. It's grown up literally from one small cell and it's produced sixteen tennis court sizes of product . . . But at the end of the day you are receiving somebody else's cells, that's the ethical issue when you really stop to think about it. Or somebody else's bit of the heart is sitting in you. That's what it comes down to. (R2)

A number of points can be noted right away: first, the source of cells is infant (neonatal) foreskin from Jewish male babies. The cells are genetically male. Both descriptions of the product from these male interviewees draw on sporting metaphors – enough skin to cover a football pitch or several tennis courts. The mother's medical history and blood are the focus of the screening to assess the safety and quality of the source material, and, as part of these checks, she has to give an account of the paternity of the child. Apligraf is used mainly in the treatment of venous and diabetic ulcers of the leg. In company literature and reports of the testing and clinical trials of Apligraf, more women than men received the product (e.g. Eaglstein *et al.* 1999), and, in one study of wound complications following coronary-artery surgery, which can be treated using Apligraf, simply 'female sex', being female, is noted as a risk factor (Allie *et al.* 2004). Other industry-funded studies of the technology don't report on the sex of patients enrolled in the studies (Dolynchuk *et al.* 1999), but a systematic review of leg ulceration suggests that more women than men have leg ulceration, and it is more common in older people (Graham *et al.* 2003). Although no systematic data on the characteristics of approximately 250,000 users of Apligraf are available, one estimate is that more men than women have diabetic foot ulcers, a group that is younger than those with venous leg ulcers, who are predominantly women.[1] The masculine imagery is striking – the relatively immature, but unavoidably masculine, characteristics of cells from penile foreskin collected from male babies being expanded and engineered to produce sheets of skin that, in scale and size, could cover a football pitch or several tennis courts offer a powerful symbolism of male potency, a masculinist dream. And the skin produced is most likely to be applied to the body of an older woman, whose ulcerated leg may be chronically affecting her ability to lead a fulfilled and pain-free life.

Another fieldwork experience affords a very different set of images. At the first conference I attended on SCs in 2004, at the University of Warwick, I was struck by the large number of women present compared with conferences I usually attended.[2] Most were bioscientists (only three social scientists were present). Then, during one of the breaks, I was introduced to a very well-known SC scientist who described his work as 'more like cultivation' than science and talked about the large number of women he had working for him in his laboratory, because they were good at 'nurturing cells' and looking after 'his babies', that is, the embryos/blastocysts from which he had successfully derived SCs (field diary, 26 July 2004). I found such talk shocking, and it left me with a lot of questions about just how SC science was gendered, how such an account implied a highly gendered division of labour in the laboratory setting, and whether this went some way to explain why a large proportion of women were present at this conference. By comparison, I had been attending conferences that were focused on issues around the emergence of TE as a field, the problems of commercialization and regulatory obstacles to industry development. Men dominated these events and later discussions I observed, when the debate extended to commercialization of 'regenerative medicine', the appropriate business models for companies and the need for manufacturing technologies, especially bioreactors, that could 'scale up' production, industrialize laboratory practices and assist in the translation of new therapies into the clinic. Such observations are indicative of a gendered division of labour within the broad field of 'regenerative medicine' and significant distinctions between the practices of TE and SC science.

Engineering is, after all, well recognized as a highly gendered activity, and substantial evidence for the under-representation of women in all areas of engineering has been highlighted by feminist scholars and initiatives to attract girls and women to science, engineering and technology.[3] TE as an interdisciplinary field has brought together chemical engineers, material scientists, biologists and biochemists/pharmacologists, but a trawl through the lists of directors of major TE laboratories and TE company profiles revealed a much higher proportion of men leading them. Gail Naughton, who trained as a biologist (Advanced Tissue Sciences, US), and Sheila MacNeil, a physiologist and now Professor of TE (Celltran, UK), stand out as two women who are prominent in the field of engineering skin. Professor Dame Julia Polak, who trained as an endocrine pathologist (Imperial College, Novathera, UK), is both a distinguished scientist and the recipient of a heart and lung transplant in 1995 and is one of the longest-living heart–lung transplant survivors in the United Kingdom. Her experience as a transplant *patient* is foregrounded in accounts of her prominence in the field of regenerative medicine. Commonly, however, it is men who dominate the discussions about commercialization, business models and production methods. In our study of TE between 2002 and 2004, where we sought to identify leaders in the field – scientists, clinicians, manufacturers, policymakers, consumer groups and other stakeholders in the United Kingdom and Europe – most of our interviewees were men. Only twelve of the seventy-three interviews were with women, and, of these, three representing consumer groups were health professionals (two nurses and a podiatrist), one worked in the purchasing agency of the NHS, one was a medical director of a national transplantation agency, four worked in industry in regulatory affairs or policy, one had a background in biology and was head of medicine in a company, and two were academic scientists. Conscious of the underrepresentation of women in the field, we sought out opportunities to include women in our sample, but the outcome was a strong indicator of the finding that men were shaping the field, developing the science, technology, clinical practice and policy.

Women are more likely to be found in SC laboratories doing 'wet biology' and the routinized, detailed work of cell culturing and 'nurturing', working as postdocs, research assistants and researchers, while men lead the research teams and laboratories. In all areas of science, few women reach senior research positions, direct research groups or centres/institutes or become professors, although women have attained international recognition in reproductive biology: for example, Anne McLaren (Cambridge) and Alison Murdoch (Newcastle) are both significant figures in the field of reproductive technologies and SC science.[4] Horizontal and vertical segregation in science has become a policy action area for the EC, but evidence of gender inequality persists.[5] Although proportionately fewer than men, there are more women found in the life and medical sciences than in engineering and technology. So what might be the consequences of this gendered workforce for how regenerative medicine develops?

Although science and technology studies have shown how sociotechnical networks co-construct technologies and social relations, they have often neglected gender relations:

> Most scholars are habituated to considering gender issues only when their subjects are women. Mainstream studies (of science and technology) have generally assumed that gender has little bearing on the development of technology because the masculinity of the actors involved was not made explicit. Despite a burgeoning literature on men and masculinities, the

critical role played by technology in hegemonic masculinity has been largely ignored. It might be seen as ironic that the focus on agency has rarely sensitized these authors to issues of gendered subjectivity. By bracketing issues of sexual difference and inequality, mainstream technology studies fail to explore how technologies operate as a site for the production of gendered knowledge and knowledge of gender.

(Wacjman 2004:45)

Wacjman argues that 'technofeminism' entails a re-visioning of network analysis to reveal how gendering of the design, innovation, manufacturing process and use of technologies occurs. In other words, how gender is produced and constitutive of all aspects of technoscience. A feminist approach implies raising concerns about women's access to technology, the impact of technologies on women's lives and the patriarchal design of technologies, but, according to her, should also attend to how 'artefacts embody the relations that went into their making', and

that the stabilization and standardization of technological systems necessarily involve negating the experience of those who are not standard. Networks (of actors and artefacts) create not merely insiders, but also outsiders, the partially enrolled, and those who refuse to be enrolled.

(ibid.:42)

It follows that, in order to develop a feminist understanding of tissue and cell therapies, we need to explore how gender shapes (and is shaped by) the scientific and technical practices of design, innovation and technology development, but also the symbolic meanings attached to, and encoded within, technologies. Drawing on Cockburn and Ormrod's study of the microwave, Wacjman explains how

gendering does not begin and end with design and manufacturing. Domestic technologies are also encoded with gendered meanings during their marketing, retailing and appropriation by users. Whilst the technology is made into a physical object during production, the symbolic meanings attaching to it are continually being negotiated and reinvented.

(ibid.:47)

Technology users are configured in specific ways, although these may be renegotiated and reinterpreted. The notion of 'interpretative flexibility', a concept first introduced by Pinch and Bijker, points to the ways in which technologies evolve in their implementation and use (ibid.:37), how they are flexible within certain limits.

So, what kind of masculinist imaginary is symbolized by the football pitch of skin produced from penile foreskin? What kind of masculinity is expressed by the SC scientist recounting how women workers in the laboratory are well suited to the work of 'nurturing' embryonic SCs? And how do these constructions position women in TE, SC science and regenerative medicine?

Defining clinical (social) need

Regenerative medicine is often described in terms of its potential to treat the problems of an ageing population and to address a shortage in the availability of organs for transplantation. UK Transplant data reveal that the majority of recipients of organ transplants are men. There are also gender differences among donor groups – for example, more, younger men, compared with women under thirty-five, donated kidneys (2006–7), and more women over the age of fifty, compared with men of the same age, became donors. Hence, for example, in 2006–7, 52 per cent of kidney donors were men, and 61 per cent of recipients were men, compared with 48 per cent women donors and 39 per cent women recipients. For pancreas transplants, 55 per cent of donors and 62 per cent of recipients were men, compared with 45 per cent and 38 per cent of women in each group. Most cardiothoracic transplant recipients are also men – 62 per cent – compared with 38 per cent women, and, for liver transplants, 58 per cent of recipients are men, and 42 per cent are women.[6] In each category, therefore, more men receive transplanted organs, and the difference between the number of male and female recipients is greater than the difference between the number of male and female donors. More men are on the transplant lists for each organ. Most organ recipients are under sixty years of age. Transplant data for the United States show similar dominance in the numbers of men.[7] Therefore, even if it becomes possible to 'engineer' whole organs (which many commentators now argue is unlikely), this suggests that more men than women would be likely to benefit from such technology. Women, however, might benefit from better treatment for degenerative diseases. Many of the targets for current SC research are neurodegenerative diseases such as HD, PD and Alzheimer's disease. HD affects both men and women equally. Men are slightly more likely to suffer from PD, but women make up a larger proportion of those with Alzheimer's disease or other dementias, because they live longer.[8] Neurological disorders caused by trauma – stroke and spinal-cord injury – are also a target for new therapies. More men suffer spinal-cord injury from falls, accidents or violent incidents, and many are relatively young.[9] Diabetes, another target disease, affects slightly more men than women. According to national UK data for 2004–5, 51 per cent of those with diabetes are men and 49 per cent are women, and this is a consistent pattern over time.[10] Many sufferers of Type 1 diabetes are children and young people.

In Western societies, women still live longer than men and therefore constitute a larger proportion of older people. However, feminists have highlighted gender differences in patterns of morbidity. Women's experiences of health and illness have been understood as distinct from men's, and such differences have been explained in terms of their different life chances, access to material resources, lifestyles and social roles, as well as the politics of reproduction (Wilkinson and Kitzinger 1994, Sabo and Gordon 1995, Annandale and Hunt 2000). Additionally, feminist analysis has indicated that clinical and professional practices are often discriminatory and perpetuate gender stereotypes and inequality, and that drug development seldom takes account of the distinctiveness of female physiology (for example, the interaction of the menstrual cycle with drug regimes). Although some studies essentialize and naturalize women's bodies, there is substantial evidence of important gender differences relating to the aetiology of disease, the incidence of disease, trauma injuries and the experience of health and illness.[11] This implies that a closer reading of the emergence of innovative health technologies, including tissue- and cell-based therapies, is necessary to assess the implications for different social groups, and in particular men and women. Such an assessment, I suggest, should include examining carefully the specific diseases and health problems being targeted by 'regenerative medicine'. This is a much bigger project, and here I offer only a preliminary and partial look at limited data in my investigation of the emergence of tissue-engineered products. By 2008, only skin and cartilage tissue-engineered products had reached the market place.

Cartilage repair and regeneration

The development of autologous chondrocyte implantation (ACI), which I explore in more depth in the next chapter (and see Kent et al. 2006), has been primarily targeted at patients with knee-trauma injury. It is a technique for repairing cartilage lesions. Such injuries, often sustained through sporting activity, are more likely to affect men. From analysis of the clinical literature, interviews and industry publications, what emerges is a view of ACI as of limited availability, usually via private clinics, to 'generally young and healthy individuals'.[12] Moreover, a complex and extensive programme of post-operative rehabilitation is required to promote healing and a return to full knee function (Bentley and Minas 2000, Hambly et al. 2006).[13] Images of young, fit men cycling, rowing, running and playing football dominate the literature, and soccer players have been a focus for some studies. In one study, which reported on findings from a sample of both men and women soccer players, 29 per cent of the sample were women (Mithofer et al. 2005). It was argued that osteoarthritis, a common condition in the joints of older people, is not a suitable target for ACI, as one orthopaedic surgeon explained in an interview in 2003:

A: The patient groups are the, they're young, obviously between the ages of about thirteen and about fifty, and they're patients who've essentially got an injury to the articular surface of the joint but who otherwise have a normal joint, so they tend to be young sports men and women or people who have experienced a trauma.

Q: So, there are, is there a demand coming from them for new procedures of this type?

A: Yes [. . .] Well what happens is that they, it's now, of course, been advertised a lot in newspapers and so on and radio and television what not, but the – so it's fairly well known amongst the athletic fraternity [. . .] I think most of our patients are pretty athletic, they're young, active people, but only maybe, 20 per cent are what I would call top class athletes, a lot of them are recreational athletes who are quite good . . . and, so we have confined it to that and to that group of patients. But there is a big question about whether you can use it for osteoarthritis, and we have done a few patients, who clearly had signs of osteoarthritis, not very advanced, and we don't know the results yet actually, they're not out at the moment but, I mean, it would be wonderful if you could use it early on but as you're aware the joint contains a lot – as soon as the cartilage breaks down to any extent then, of course, it will release a lot of enzymes from the cartilage [. . .] so we don't know whether that's ever going to be a possibility but if it were, it would be, of course, widen the scope of the application. As it is, we did a sort of straw poll of the members of the British Association of Surgery of the Knee and based loosely on that we reckon there might be ten thousand patients per year in the UK in this group who would be appropriate for the treatment. The way they present is with pain. You know, they've had a sports injury or some other injury and they present with the pain, sometimes swelling because of the effusion and all we find is that they've lost a piece of the articular cartilage. (CL5)

Another clinician was highly critical of the lack of clinical trials testing ACI and failure to address the needs of this much larger group of patients:

Well [. . .] it's a duff idea for osteoarthritis. There is one indication where it just possibly might work but I doubt it and that's where you got an isolated area of traumatic damage to the articular cartilage in a relatively young person, where it is possible that you could produce a repair of that defect. However, in the vast majority of cases and certainly in all the osteoarthritis, the problem of cartilage damage is induced by abnormal biomechanics of the joints, so the joint's got funny loads going through it and that causes the damage. So to try and repair the damage without correcting the biomechanics is just fundamentally stupid, I think, and I can't really, genuinely can't understand why academics and pharmaceutical companies are giving so much of their time because it seems to me to be fatally flawed by that rather obvious fact, that this is all driven by abnormal biomechanics and so if you repair the defect it will just fall to bits again because you've still got the abnormal biomechanic, so I think it's just all crap. I even think it's probably crap to go for the repair of the single defect in a younger person because there's actually no evidence that there's much point in doing that in the first place, and one of the huge flaws in my view, of all of the research that's been done so far, is the, again to me, crashingly fatal flaw, that we don't know the natural history of the lesion left alone, and there's quite a lot of evidence of the leaving alone and they do just fine, thanks very much – that without any treatment at all. But this is a thing that's been driven by money in companies and by the ego of individual surgeons, and none of those are interested in the natural history of the defects, so I'm extremely sceptical of this whole approach. (CL1)

His view, then, was that developments in the field were being shaped more by the incentives and egoism of colleagues, who were motivated to intervene in ways that had no proven benefit compared with existing treatments.

The drivers are, a lot of money potentially and a huge ego, a large number of surgeons who just have to do something, if they see something wrong with the joint and I fear that those drivers will overtake common sense or data and I think it will be, well one of the outstandingly big problems in this whole field and area is that it's very, very difficult to get the good long-term data on outcomes that we need, so the classic in this is with joint replacement, where we're always inventing a new joint replacement before we know whether the last one was any good or not, because it takes you about twenty years to find out whether a new design is any good. People just can't stop tinkering and the, the all time classic, and this is well known and well published now, is that of hip replacements, Charnley hip replacement, nobody has ever, ever done better than Sir John Charnley did originally in his first prosthesis in the 1960s, but we have something like sixty, you know, better than this, you did review this yourself, Alex, but you got something like fifty designs now, of hip replacements, but they're not better than Charnleys, and so why we've done all that? (CL1)

Patients were reported as seeking out ACI, but many were unsuitable to have the treatment, especially older groups:

Q: Are you, you're finding patients increasingly well informed about possibilities?

A: Yes, yes, I mean, increasingly they'll come in with an article from a newspaper or sometimes they look at the Internet, you know. They say, I've read about this, what about this, so yes and particularly in the sporting fraternity, you know, where's it obviously very important to get it fixed and recover as soon as possible. [. . .] I mean, we've had an association with the Football Association, it's a procedure we do on soccer players. [. . .]

Q: Partly just out of curiosity, I mean, are there particular sports, which are more prone to this particular condition?

A: Yeah, soccer, yeah and any sport involved impact and rotation of the lower limbs, but soccer and rugby football are the main ones and there are quite a lot of people who, and tennis is actually but not many people because tennis isn't a big game like football and rugby.

Q: And that is, is it important to the development of the market for these, as it were, that there are some presumably relatively well-known, you know, people, in the pubic eye who are . . .

A: Yes, I think so, yes. I mean, by far the greatest number of enquiries, of course, come from older people.

Q: In older people?

A: Yes, who've got established arthritis and they read that you can replace the cartilage in the joints and they are sensible enough to know that the cartilage is broken down and they say, can I have some new cartilage? Almost always it's not suitable because the joint is not bio-mechanically reliable, there's deformity there and excessive loading of that side of the joint, of that cartilage, whichever. And also osteoarthritis has an inflammatory component as well, so it's one of the degenerative components, but by far in a way the greatest number of communication is by letters and so from patients who are older, and a lot of the youngest don't know that about it, unless somebody tells them and they're not so you know, not so interested in reading about medical matters and so on. (CL5)

ACI was pioneered by a Swedish orthopaedic surgeon, Lars Peterson, who, from his work in the 1970s, identified young injured athletes and active patients as 'problems', because, at that time, there were no treatments for articular cartilage lesions (Brittberg *et al.* 1994, Peterson 1996, Peterson *et al.* 2000, Peterson *et al.* 2002, Lindahl *et al.* 2003). He was influenced by reading an earlier paper by an English woman scientist, Otis Smith, who was the first person to isolate articular chondrocytes and to grow them in culture. By 2003, an estimated 20,000 patients worldwide had had ACI procedures, most of them in the United States (interview with CL-EU3). Peterson was enrolled by Genzyme, an industry leader, to train surgeons in ACI in the United States and host training events in Sweden from 1998. Networks between commercial providers and clinical scientists were important to the development of the technology and central to discussions about scaling up production and moving from a custom-made autologous product (see next chapter) to an allogeniec product (for a discussion of bionetworking and recruitment for experimental therapies in India, see Patra and Sleeboom-Faulkner 2009).

Q: Yes, you were talking about access, the users, and reimbursement. Do you think [. . .] this is an expensive treatment?

A: Yes but I think if you have a more defined regulatory system you will be able to produce this in a larger scale and if you have acknowledged laboratories they can do a larger production and then the prices, probably, will go down. It's really, it's really a high-technology system and high safety demands, so you have to pay for it but still I think, with competition from established laboratories, the price will go down.

Q: But you were also talking about autologous treatment, so it's a fairly patient-tailored, customized treatment.

A: Absolutely, absolutely.

Q: You would need a lot of lab [inaudible].

A: So, but the development may be going into allogeneic treatment or the articular embryonic stems that's made or mesenchymal stem cells maybe in the future.

Q: Do you think that it's, it would be commercially attractive to companies as long as it's only autologous material and not donor material? Do you think it's more commercial?

A: I think, I think as long as it's autologous material it's always a complex logistics and that costs money, and it's also very time dependent. You have to do it from that time you received the cells and you have to deliver it at the time of surgery, so it's really complicated. Allogeneic cells you wouldn't have that problem, you can produce the cells and grow them. (CL-EU3)

Scaling up here is seen as both reducing costs and widening access to treatment and as a necessary precondition for better reimbursement models within health-care systems. Peterson and Lindahl, his collaborator, set up their own company, Cell Matrix, in 2001, and obtained a licence from Fidia Advanced Biopolymers (FAB) (an Italian company) for use of its scaffold delivery system, Hyalograft C,[14] in Scandinavia. FAB also worked in collaboration with a scientist in the United Kingdom (Pavesio et al. 2003) and subsequently co-ordinated a Europe-wide project to develop a systems approach to TE processes and products (STEPS).[15] The use of a scaffold on which cells were seeded was seen to have benefits over the technique of creating a periosteal flap in the joint under which cells were implanted. Other 'matrix-assisted' approaches include a Belgian company, Verigen, which uses a collagen matrix (MACI).

Although initial, 'first-generation' ACI targeted sports injuries and was shown to be successful in patients with osteochondritis dissecans (OCD), its application for the treatment of osteoarthritis has been a focus of further study and investigation (Hollander et al. 2006, Richter 2007). Osteoarthritis is more common in women and in middle and older age. It is the most common joint condition worldwide.[16] A primary reason for implanting patients with osteoarthritis in the knee would be to delay (or prevent) the need for a knee replacement, a procedure that has its own limitations. Since 2000, in the United Kingdom, the National Institute for Health and Clinical Excellence

(NICE) has not supported the use of ACI in osteoarthritic joints, nor as a treatment of choice, but only as part of clinical trials – a view that persisted in the absence of any additional clinical data in its reviews of evidence in 2005 and 2008 (National Institute for Health and Clinical Excellence 2000, 2005). In an interview in 2003, this scientist agreed that the clinical data were inadequate:

> If you look through the literature for Autologous chondrocyte implantation, the quality of outcome data in terms of analysing the tissues is very poor – it's basically just histology pictures – there's a lot of clinical data in terms of asking the patients how well they're doing and arthroscopic assessment of what things look like when you put a camera in the knee. But when people have been taking biopsies and studying them all they've been doing is cutting histological sections and looking at the structure in that way and we've invested a lot of time and effort in developing sensitive biochemical techniques which allow us to ask a lot more questions in a more objective way because we're measuring constantly the components that we would normally expect to be in cartilage and putting that together with histology and other outcome measures. And thinking in regulatory terms I think this is a real issue because its very easy for people to show a nice picture of what looks like a cartilage and say we've got a great product here, and that's not good enough, that's why I've been pleased to work with FIDIA because they've been very open minded and they've allowed me to look at their biopsies using these newly developing techniques. And even though it's not shown perfect results for their product by any means, they're happy to work with us on publishing this and showing this as a benchmark of what can be done, what we can look at. So over the next few years we want to consolidate that, look much harder at patient outcome in the longer term, OK. I think incidentally there needs to be a lot more effort in developing better clinical outcome measures; my belief is the clinical pool being used to assess how patients are doing is not sensitive enough; which is why in virtually every technique you look at for cartilage repair you get something like an 80 per cent improvement rate across the board. I think that basically tells you there's a very blunt tool being used to assess outcome. So that's the next few years and then during that time I would like to see the best of these techniques – of which I would certainly think Hyalograft C is one, becoming available much more widely, assuming that the early data support its use. The way I would like to see that go is that there are regional centres such as this one delivering that sort of technique for the South West Region, the North, the Midlands or whatever. I know Fidia would like to control everything from where they are, I think for the UK that's not realistic. I suspect the only way to make a big market is to have local delivery areas. (S4)

What is described here is both the development of the technology and of the methods and tools to evaluate its efficacy and attempts to develop an effective business model for delivering the treatment to the clinical setting. If ACI use could be successful for patients with osteoarthritis, the predicted demand and market for the technology would be large, and, rather than using autologous cells and providing local or regional cell-culturing services, the idea that an 'off-the shelf' product would be needed was frequently articulated. This same scientist described some of the technical difficulties in evaluating these new technologies and the limitations, in his view, of autologous cells:

> However, there are problems, there are multiple problems, with using the autologous knee cells, one of which is if you want to tackle older patients with primary osteoarthritis rather than young patients with a disease secondary to injury or trauma, their knee cells – the chondrocytes in their knees – are probably not very good at making new cartilage, probably not the right cells for us, and indeed we found that adult cartilage from the knee is really not good at engineering full cartilage in the test tube. If we try and grow a proper piece of cartilage rather than just implanting cells it's very hard to do that with those, and so some years ago – three or four years ago – we identified that cells from the cartilage in the nose were somewhat better. And so we're looking at nasal chondrocyte sites and actually that's an interesting regulatory issue, because how far do we have to go in terms of experimentation before convincing ourselves and patients and orthopaedic surgeons that we should be taking cells from the nose and putting them in the knee? There's a lot of knee-jerk resistance to implanting noses in knees, but actually we think the cells would probably be better. (S4)

So, he was looking at sourcing cells for implantation from the nose and also bone marrow for treating cartilage defects and believed that there were potential applications that could treat joints with osteoarthritis. As I discuss in more detail later, the business model for autologous cells and that for allogeneic product are seen as very different and relate to the patient group that might use the technology (older people with knee osteoarthritis or younger sports/trauma patients). Moreover, the criteria for defining acceptability of this technology (as with other 'high-risk' technologies, such as gene therapy) were being constructed:

> When we're dealing with a disease like arthritis which is not terminal, although it has a significant negative effect on the quality of life, your margin of safety in your therapy whatever it is – drugs or whatever – has to be that much higher, so gene therapy already in clinical use I think raises significant safety questions and then for a non terminal disease that safety margin has to be even higher. (S4)

The STEPS project aimed to develop 'a systems approach to the design, manufacture and delivery of practical, ethical, viable and affordable TE processes and products'.[17]

In summary, the example of ACI is suggestive, illustrating how clinical problems and need have been defined and particular ways of envisioning users are being engineered into the technologies. On the one hand, the early use of ACI sought to return healthy, young, mostly male, patients to activity and a high level of functionality in the joint. On the other hand, the older patient with osteoarthritis was construed as in need of a solution that would relieve pain and delay or prevent knee-joint replacement. In a discussion of the 'skin business' in Chapter 1, we saw that the development of a market for skin products has also been patchy and limited. The debilitating effects of chronic ulceration, especially in older age (more common in women; Graham *et al.* 2003), or other hard-to-heal wounds, with associated risks of infection, immobility, pain and expensive nursing care, were drivers for innovators in this field. Few burns patients (many of whom are children) are likely to have received 'engineered' skin. Only an estimated 600 patients had received Genzyme's product, Epicel. In 2008, Celltran were conducting trials for the use of Myskin in burns treatment, and Organogenesis had still not embarked on full-scale clinical trials for the application of Apligraf in burns management.

Women's labour: gendering the bioeconomy

So far, I have suggested that women represent certain types of worker within the emerging field of regenerative medicine (and the production of new scientific knowledge). In the fields of TE and technology transfer, women appear underrepresented. In the field of SC science, they are less likely to be leading research laboratories, although the numbers of women in the biological sciences is higher than in engineering and other areas of science and technology. Women are also more likely to be found in the clinical areas working as nurses, physiotherapists or carers, rather than leading clinical innovations or developing new surgical techniques. I have also suggested that women may, or may not, benefit from tissue- and cell-based therapies in ways that are different from the ways men do. It seems likely that the target diseases and applications of, for example, tissue-engineered products have different effects for men and women. This isn't necessarily a matter of essentializing gender differences, but rather pointing to the ways in which problems are defined and selected, and how technologies configure different types of user. Moreover, women's bodies and reproductive capacity are being mobilized in specific ways in attempts to advance regenerative medicine.

Waldby and Cooper forcefully argue for recognition of women's reproductive and *clinical labour* within post-Fordist biotechnology industries, claiming that, many of the new technologies associated with regenerative medicine – embryonic stem cell research, savior siblings, somatic cell nuclear transfer (SCNT), cord blood banking – rely on female reproductive biology as a generative site. [. . .] These technologies effectively convert the generative power of female biology into regenerative therapy. Hence, they position reproductive biology as one of the most important machines for the bioeconomy – especially as a promissory machine, working through appeals to biological potential and the future regeneration of the body.

(Waldby and Cooper 2008:58)

Taking the example of a global trade in oocytes, according to them, new forms of female reproductive labour have emerged in the global bioeconomies and can be seen as a particularly feminized form of *clinical labour*. Clinical labour is understood here as 'the biological labour of living tissues and reproductive processes', which is made available for exploitation and creation of surplus value. Drawing parallels with the outsourcing of clinical trials by drug companies to poor populations in the developing world and the (often illegal) selling of organs and other body tissues, they suggest that tissue suppliers and experimental subjects are essential to the biotechnology enterprise, and that a sexual division of labour characterizes the dynamics of a neoliberal political economy. According to their analysis, older forms of gendered labour are being replaced (or at least supplemented) by newer forms of exploitation, a new biopolitics. This was most starkly highlighted by the international furore that broke out in 2006 when Korean scientist Hwang was charged with scientific fraud and exploiting the women working in his laboratory by harvesting eggs from them for research work (Kitzinger 2008, Waldby 2008, Gottweis and Kim 2010). Not only were the women workers in the laboratory labouring as part of the research team, but their contribution was extended to include the use of their body parts – ova – for the national research effort. The eggs were enucleated for nuclear transfer, so-called somatic cell nuclear transfer (SCNT), or 'therapeutic cloning'. SCNT is a technique that requires a large number of ova to increase the chances of success and, as it turned out, was highly inefficient and unsuccessful.

To date, attention has been increasingly focused on women as 'donors' and suppliers of the biological materials that are being used – ova, embryos, fetal tissue and umbilical cord blood – in SC science and the development of regenerative medicine. In relation to ova,

it is where ARTs connect up with the market in eggs that reproductive medicine in general taps into a less familial, more savage kind of reproductive labor. In the sense that egg markets are increasingly drawing on the underpaid, unregulated labour of various female underclasses, the differences between human reproductive medicine and the brute commodification of labor and tissues that prevails in the agricultural industry becomes difficult to maintain. Moreover the rise of transactional reproductive work demands that we rethink some of the key assumptions of feminist bioethics, displacing the salient questions from the realm of care, dignity, respect, and the liberal ethical contract (of informed consent) to that of labour relations and unequal exchange. This was a move first made by feminist theorists in the context of the first world, Fordist family wage. Now more than ever, it urgently needs to be reconsidered on a transnational level, in the same way that postcolonial feminists have begun to look at the global dynamics of sex work.

(Cooper 2008:136)

The international trade in eggs for reproductive medicine positions women as egg sellers, and, in diverse circumstances, they are enrolled in nation-building and reproducing national identities (Nahman 2006, Gottweis and Kim 2010). International opposition to the harvesting of women's eggs has been mobilized, for example, in the 'Hands off our ovaries' campaign[18] and, as we saw in the last chapter, has been controversial in the United Kingdom but was recently permitted for research purposes. Waldby (2008), Cooper (2008) and Franklin (2006a) explore the interface between IVF and SC science and the traffic in reproductive tissue in these new economies. Focusing on the US market, Cooper argues that, 'reproductive and regenerative medicine participate in different kinds of "tissue economies" or rather function as different phases or moments of the one, highly stratified economy' (Cooper 2008:130). For Cooper, there are important fundamental differences between reproductive and regenerative medicine. These differences relate, first, to the different notions of 'generativity' that each draw on and to the different ways in which value is accumulated.

Reproductive medicine can be seen as emerging from a historical context of Fordist production methods in the US cattle industry (Franklin (2007a) explores the links between human reproductive medicine, animal husbandry and the British sheep industry), although Cooper argues that, despite shared techniques, reproductive medicine in the United States has taken the form of a highly privatized domestic service, reincorporating the realm of social, sexual and biological reproduction within the economic sphere in a way that is more post-Fordist. Moreover, it is the productive power of reproduction itself, rather than the tissues, that has value added. In the context of the strict regulation and protection of human embryos, the institutional arrangements for reproductive medicine in the United States prevent them circulating as commodities. However, in egg markets, the laws of the market prevail, and the egg trade is relatively unregulated. Cooper argues that ova are distinct from embryos in so far as they have become tradable commodities in the United States.

By tracing cell theory and shifting ideas of 'generativity', Cooper provides us with illuminating insights into the contrasting notions of value attached to embryonic bodies in SC science and regenerative medicine. Contrasting the Weismannian theory of generation, which is where bodies are seen as generated by vertical transmission and hereditary information through the germ line, with recent SC science, we see how the potentiality and plasticity of embryonic SCs are themselves a surplus of life: 'What stem cell science seeks to produce is not the potential organism – nor even this or that particular type of differentiated cell – but rather *biological promise itself, in a state of nascent transformability*' (Cooper 2008:140).

Hence, whereas reproductive medicine perpetuates the production of the germ line, SC science more radically undermines and undercuts it. Moreover, in the accumulation of value, what are traded are 'embryonic futures', as part of a broader trend towards futures trading in the United States and world capital markets. This market rests on new forms of property – biological patents and intellectual property – that go beyond, and are qualitatively distinct from, traditional commodity markets. Therefore, she says that US patent law, since 2001, has incorporated a generic process of human (re)generation into the laws of invention, while leaving the biological person in tact (ibid.:147). It is the ownership of the speculative value of the cell line, the intellectual property, rather than the line itself that is most important in this restructuring of the bioeconomy. Hence, she argues that this precedes the process of commodification and overwrites property claims to the biological product itself.[19]

Two points are central, then, to gendering the bioeconomy: it is important to recognize, first, that the capitalist system of exchange is being transformed, and, second, that women's position within it is tied to their reproductive work. As Cooper and Waldby so eloquently argue, the procurement of ova, embryos and umbilical-cord blood is central to efforts commercially to exploit the life sciences, and this relies on a particular form of feminized labour (Waldby, 2002a, 2006, Waldby and Mitchell 2006, Waldby and Cooper 2008). The promissory and speculative nature of the capital being produced does not weaken the claim that women's bodies are being commodified, but rather points to, as I see it, an intensification of that process.

The fetal-tissue economy

American political culture and the position of the United States within the global economy elaborated by Melinda Cooper provide a particular framing of the politics of regenerative medicine. Linkages between neoconservatism, neoliberal economics and pro-life or the culture of life are drawn together in her analysis of US policy on SC science and regenerative medicine. This prompts me to ask a number of questions about how we might understand other dimensions of the gendered bioeconomy and different tissue economies or different strata within the tissue economy. Specifically, my research looking at the fetal-tissue economy and the use of fetal tissue in SC science in Britain highlights overlapping but distinctive features when contrasted with 'embryonic economies' (or oocyte markets) (Kent 2008a).

There are a number of key features of the fetal-tissue economy in Britain and a number of comparisons that can be drawn between the fetal-tissue economy and what Sarah Franklin calls 'embryonic economies' (Franklin 2006a). First, abortion is a common procedure, strictly regulated by law, but abortion *in order* to provide fetal tissue for research is illegal.[20] It is permitted for fetal tissue to be used in research, and there are guidelines for researchers, known as the Polkinghorne guidelines (Polkinghorne 1989), but these have been interpreted in different ways, and there was local variation in the kinds of information given to women asked to donate tissue to SC science. There has been no national oversight of research using fetal tissue, and, outside the scientific community, little was known about the numbers of fetuses collected or their value to SC science (Pfeffer and Kent 2006, 2007). The extent to which women should be given specific information about the use of the tissue in research was a matter of concern and confusion among scientists in this research, who thought that the Polkinghorne requirement of general, that is, non-specific consent contradicted good practice in other areas, such as the collection of embryos for SC research. The notion of informed consent, therefore, was problematic in this context, but there were also suggestions that some women might not want to know what happened to the fetus after the termination. Research nurses acted as intermediaries between the abortion clinic and the researchers in the laboratory. In contrast, as the work of Franklin and others testifies, there have been concerted attempts to develop national standards for hESC derivation and banking and ethical protocols for patient information and informed consent (Franklin *et al.* 2008). The supply of embryos for SC research in the United Kingdom has been more strictly regulated and centrally managed.

Second, women who have an abortion and those undergoing IVF treatment may be seen as different kinds of donor. In this study, despite the fact that abortion is a common procedure for women from all backgrounds, the former were sometimes perceived as more likely to be engaged in 'risky' (sexual) behaviour and to produce material that is both biologically and symbolically contaminated. Donation was thought to be a way of redeeming them as good citizens, but also potentially to assist with grieving about the termination (see also Hogle 1999, Ariss 2003). There was no evidence of expectations that research would benefit women undergoing abortion, nor of reciprocal exchange relationships, such as those in IVF clinics described by Franklin (Franklin 2006a). However, although it was suggested that, in contrast to women in IVF clinics, women undergoing abortion had no long-term commitment to either abortion care or research, the reasons why some women will donate fetal tissue were largely unexplained by the data collected in this study.

Third, a woman's relationship to the aborted fetus was seen as different in important respects to that of the couple in the IVF clinic making decisions about the fate of their embryos. In Britain, legally, the fetus is the woman's tissue, or 'mother's tissue' – a part of her body (Human Tissue Authority 2006). It has no separate legal status, but, despite this, could be viewed as a corpse deserving of the respect for a dead person and disposed of by cremation or burial, and tissue collection was sometimes seen as like organ donation. Specific practices of 'sensitive disposal', which took place in some, though not the majority of, abortion clinics, were linked to particular notions of respect for the fetus as having a special status (Royal College of Nursing 2002). Although most abortions take place in the independent sector, the SC scientists interviewed had local arrangements with NHS hospitals in the public sector to collect tissue. Most collected tissue following surgical abortion, although medical terminations can produce a fetus that is intact, which has some technical advantages. At the same time, the notion of fetal remains as 'clinical waste' shaped policy and practices relating to disposal of tissue in the hospital and laboratory by incineration. Although the pre-implantation embryo is afforded a high level of protection under British law, the dead fetus both has a different legal status and is assigned different meanings, which in turn shaped practices in the abortion clinic and the laboratory. Both the context and the character of the exchange relationships between women donating fetal tissue in the abortion clinic and those procuring and using the tissue are distinct from those relationships that characterize the 'IVF–stem cell interface' (Franklin 2006a, Kent 2008a).

Fourth, the fetus is a source of diverse cell types and has a range of potential uses and was regarded as a highly valued biological resource (as are pre-implantation embryos). The discourse of 'waste' provides both a justification for using the fetus and a means of releasing its latent 'biovalue' (Waldby and Mitchell 2006, Kent 2008a). The fetus has a reproductive past that is distinct from that of the pre-implantation embryo, in so far as it is a product of sexual relations. Its reproductive value, once a pregnancy is terminated, can only be defined in terms of the potential to transform tissue into regenerative cells. Women's labour in 'producing' an aborted fetus and making it available for research is also distinct from that of the woman undergoing IVF treatment. Conceiving, carrying a pregnancy even for a short period and undergoing elective termination are quite different corporeal and emotional experiences for a woman, compared with undergoing IVF treatment and having to make decisions about the future of pre-implantation embryos. I have argued, therefore, that the path from the abortion clinic to the SC laboratory is one that deserves to be better understood and taken into account in discussions about the ways in which SC science transforms the connections between death and life (Franklin and Lock 2003, Kent 2008a). Although some dead fetuses have become highly visible 'icons of life' (Morgan 2004, 2009), commonly, the practice of collecting and using fetal tissue for research is hidden from view. Mapping the fetal-tissue economy means making visible the ways in which the use of fetal tissue as a research tool in the life sciences has been commonly obscured from public view (Kent 2008b).

Hence, the social meaning and significance of fetal remains are contradictory and ambiguous. As 'waste', the human characteristics of the fetus are played down, and revaluation of its generative and productive potential is transformed by its capture within the life sciences. The transition from the abortion clinic, where a woman's sexual, reproductive history and health are a focus, to the SC laboratory is filtered by a series of practices that transform the unwanted pregnancy into a source of (stem) cells (Pfeffer 2009b). However, the potentiality of the fetus as a source of 'regeneration' is controversial in technical terms – that is, whether the cells derived from it have therapeutic potential or the capacity to become 'immortalized' or produce stable cell lines. It is also highly controversial as an exchangeable commodity. Arguably, a fetus is *more* human or person-like than an embryo and, therefore, in its acquiring biovalue, being used as a research tool or resource, the critical distinction between person and thing is undermined. An aborted fetus was likened to a cadaver in both the regulatory and policy discourse and by some interviewees in this study and, therefore, like corpses, regarded as deserving of respect and 'sensitive disposal'. However, donation of a dead fetus was also seen as similar to donating an organ. Organs made available for transplantation are both objectified and humanized, ascribed the characteristics of their donors but also produced as standardized objects for transplant medicine (Hogle 1995, 1999, Lock 2002a, 2002b, 2004). However, the use of fetal tissue in SC science in Britain was relatively non-standardized and unregulated.

There is evidence of the capitalization of fetal tissue, and a number of companies in the United Kingdom and United States have programmes that use fetal tissue in their portfolio of research and development activities; although none, so far, provides standardized treatments, some distribute lines as drug discovery tools (Kent 2008b). As we saw in Chapter 2 above, ReNeuron plc is a UK-based company using fetal-derived cells in a new clinical trial for treating stroke patients. It has licence agreements for distribution of its ReNcell 'human neural stem cell lines', via Millipore,[21] to other researchers for drug discovery (Donato *et al.* 2007) and a cross-licence agreement with Stem Cells Inc. for the production of cell lines using its expansion technology (see below).

Stem Cell Sciences plc (SCS) sourced fetal tissue from an NHS trust in Edinburgh, with the approval of an ethics committee, and women were informed that the tissue may be used in research that has a commercial application in future.[22] It had strong academic links to Austin Smith's laboratory and was conducting preclinical studies of neural SC grafts for repair of spinal-cord injuries (Daniels 2006).[23] Unlike ReNeuron, whose core business is products derived from fetal tissue, SCS had a range of research and development programmes, which included hESCs, and owned patents in both adult and embryonic SC fields. One aim was to develop SC-based therapies for currently incurable diseases. It had links with international biotechnology and pharmaceutical companies, including Pfizer, Sanofi-Aventis, GSK, Chemicon, Deltagen and Lexicon, and offices in Edinburgh and Cambridge (United Kingdom), Melbourne (Australia), Kobe (Japan) and California (United States).[24] It launched on the London Alternative Investment Market in 2005 but was sold in 2009 to Stem Cells Inc.[25]

In the United States, Stem Cells Inc. is a California-based company that produces HuCNS-SC™ – 'a cell therapy product consisting of neural cells prepared under controlled conditions. Neural stem cells are isolated from the human fetal brain, purified, expanded and then stored and frozen in cell banks until they are transplanted as HuCNS-SC doses'.[26] FDA and IRD approval was obtained in 2005–6 for an open-label phase-1 trial in children with Batten's disease (a rare neurological disease), and the first child received an implant in November 2006.[27] The company is developing studies in preclinical animal models of PD (in collaboration with the Swedish group), Alzheimer's disease, spinal-cord injury and stroke. It began a new spinal-cord clinical trial in early 2011. Neuralstem, Inc. is a biopharmaceutical company with a robust and diverse development pipeline derived from its novel, patent-protected human neural stem cell (hNSC) technology. The company's research and development efforts target the treatment of debilitating central nervous system (CNS) disorders for which no effective treatments exists.[28] These two companies were locked into a patent dispute about the technology to create neural SCs.[29] Stem Cell Innovations was a new drug discovery company, with laboratories and offices in the United States and the Netherlands. It developed research tools and produced human pluripotent germ cell lines Pluricell™ from 5–9-week-old fetuses for drug discovery and currently had seventeen banked lines. Its aims were to produce different cell types from fetal material to model different diseases, including PD and HD.[30]

In summary, then, the regulatory context and local practices relating to fetal-tissue collection for use in SC science are likely to vary across nation-states, although further research is needed to characterize and map these variations. From the study of fetal-tissue use in Britain, it was evident that women undergoing abortion are another group being enrolled in SC science, and that their clinical labour is distinct from that of those who donate eggs and embryos (and umbilical-cord blood) in important ways. One thing that feminism has taught us to recognize is that women are not equal, that, as has been pointed out in relation to egg sellers, it is more likely that poorer and disadvantaged women will be induced to sell their eggs (Nahman 2006, 2008, Dickenson 2007, Waldby 2008). Tensions have been shown to exist between a gift economy, where women may be expected to donate tissue as an act of solidarity and altruism, and a capitalist economy, which promotes the commercial exploitation of human biological materials as the starting materials for the production of new therapeutic products. Processes of capitalization lead to the creation of both surplus and promissory capital.

Conclusion

In this chapter, I have suggested that regenerative medicine is gendered in important ways. At the symbolic level, aspirations to regenerate the body through the industrialized production of body parts and cell-based products mobilized a masculinist dream of, for example, enough skin to cover a football field or numerous tennis courts. The use of male foreskin embodied an especially masculine form of potency and productivity.

The technical and practical aspects of production were seen as dominated by men, as industry leaders, senior scientists, clinicians and policymakers. There is evidence of vertical occupational segregation within the sector, with women more likely to be found in junior research positions, especially doing 'wet biology', and less likely to be found in the engineering disciplines focused on scaling up, bioreactors and production methods, although an earlier image of the automated cell-culturing system at the beginning of this chapter (a system used by ReNeuron in the production of its fetal-derived neural line) suggested that women technicians might become operators of the automated cell-production system. Innovative clinical scientists or clinicians at the 'cutting edge' of their discipline are also more likely to be men, whereas women are more likely to be working within the nursing or health-care professions, delivering long-term, chronic care. Notable exceptions that I came across were Anne Rosser (Cardiff) and Anne-Marie Bouchard-Levi (Paris), both clinical scientists with special interests in the use of cell therapies to treat patients with HD.

I also suggested that the therapeutic benefits of tissue- and cell-based therapies are likely to be different for men and women, given the evidence of differences in disease patterns and treatment regimes. This was indicated both by the higher demand for organ transplantation from men and their increased risk of trauma injuries and the higher numbers of women likely to experience degenerative diseases associated with ageing (for example, osteoarthritis and dementia).

Finally, women are more heavily implicated within the emerging bio-economy as donors, as the source of tissues and cells, as many of the technologies being developed rely on the procurement of ova, embryos, umbilical-cord blood and fetal tissue. Despite recent discoveries of techniques to produce 'induced pluripotent stem cells, redirecting and redifferentiating somatic cells', and permission in the United Kingdom to use animal ova to produce 'admixed human embryos' for research, there is a continuing demand for reproductive tissues. Indeed, it has been suggested that 'biocapital is not just dependent on reproduction, it is *constituted* by it' (Franklin and Lock 2003:10). The development of regenerative medicine is therefore an important feminist issue, as I will examine in more detail in the next two chapters.

6   Remaking the self

Introduction

So far, I have described how the development of tissue- and cell-based therapies, or tissue-engineered products, mobilizes bodily tissues in diverse ways. Once procured, tissues and cells can be more or less manipulated (for example, expanded in number or seeded on to a synthetic scaffold) and transplanted into the bodies of patients. In the future, diverse diseases or functional problems may be treated. The therapeutic aims of these therapies are wide ranging, and the types of product and approach are diverse and complex. In this chapter, I focus on the issue of what views of bodies emerge in regenerative medicine. I also want to explore how these technologies reconfigure and remake the links between bodies, identity, self and others. And I will consider how feminist bioethics might help us to think through the ethical issues at stake. I also draw on focus-group data to explore the relationship between women and fetal-tissue donation.

In 2007, another scientific development was heralded as creating new possibilities for the generation of tissues that could treat disease – the successful production of 'induced pluripotent stem' (iPS) cells. Two teams, one in Japan, the other in the United States, reported that they had successfully transformed skin cells (fibroblasts) into pluripotent SCs that had the capacity to become the cells of other tissues in the body (Takahashi *et al.* 2007, Yu *et al.* 2007). Previously, the dominant view was that, once a cell had become differentiated or specialized within the body, it could not be 'turned back' into another cell type. Until this success, only hESCs were considered 'pluripotent' and as being flexible enough to develop into the diverse cells of the body. As a result, opponents of the use of hESCs especially welcomed the news, because it seemed to obviate the need to destroy embryos for SC research. James Thomson, who headed the US laboratory, was already well known as a pioneer in SC research for his earlier work with embryos in 1998, and the *New York Times* suggested that he had now solved the ethical problems his earlier work had created, with the headline 'Man who helped start stem cell war may end it'.[1] Others insisted that both cell types were important, and that embryos would still be needed for research. So, what did it all mean? What were the implications of this latest 'scientific breakthrough'? It highlighted the flexibility and plasticity of cells from the body and called into question preceding assumptions about the generative potential of 'adult stem cells'. It seemed they could be 'reprogrammed', and skin could become heart tissue, or some other tissue type.

Underlying discussions about regenerative medicine are a number of key ideas about bodies and about how we think about or understand them. The ambition to 'regenerate bodies' draws on important ideas within the life sciences and biomedicine that merit closer consideration here:

- Commonly, bodies are construed as *natural* entities, self-determining and bounded biological systems.
- Historically, within biomedicine, the body has been viewed as *machine*-like, and some accounts of TE rearticulate this kind of perspective, in so far as body parts are seen as reparable, replaceable and *interchangeable*.
- Bodies are also understood as comprising tissue and cells, and efforts directed towards growing (engineering) body parts (or tissues) focused on culturing cells, controlling and manipulating the *dynamic* processes of biological growth.
- This, in turn, centres on notions of *multiplication and replication* – expanding the numbers of cells and reproducing cells, in a cell line, that are identical. A 'stable cell line' is one where the process of replication is consistent, without (genetic) mutation, each (stem) cell a daughter of the previous one, but genetically identical.
- Furthermore, bodies are thought of as *plastic*, flexible and malleable – and the plasticity of cells refers to the ability to control ways in which cells can be differentiated, driven down a particular developmental pathway or 'reprogrammed'.
- *Regeneration* is a core, though contested, concept that can be taken to refer to an (innate) ability of the body to repair or heal itself.

At the same time, these ideas tie in to a moral order, linking science, nature and culture.

A naturalistic view of bodies, and of illness and disease, suggests that the body is separate from social and cultural processes but part of a natural world. This view was perpetuated for many years by the idea that the study of bodies was the domain of the natural and biological sciences rather than social science. In contrast, as Lock and Farquhar (2007) show in tracing the development of historical, anthropological and sociological thought, there has more recently been recognition of the need to go 'beyond the body proper', for:

> most of social science continued for a long time to treat bodies as the naturalized, essentially passive atoms or building blocks of society . . .

the classic problematic of the relations between individual and society that still provides analytic tools for most of the social sciences seems to require a 'proper' body as the unit of individuality. This body proper, the unit that supports the individual from which societies are apparently assembled, has been treated as a skin-bounded, rights-bearing, communicating, experience-collecting, biomechanical body.

(Lock and Farquhar 2007:2)

However, others, especially feminist scholars, have demonstrated how what a body is, what it means to have a body (agency), how a body is experienced (embodiment), how a body is perceived (body image) and how bodies can be reshaped or altered are problematic and linked to processes of identification (identity) and notions of self (who I am), while the influential work of Foucault showed how bodies are disciplined, regulated and discursively produced in modern societies (Foucault 1973, Turner 1994). Therefore, rather than seeing a body as a unitary, 'skin-bounded' entity, more recently within social theory bodies and selves can be seen as multiple, relatively unstable, historically contingent, contested and malleable. Such a radical revision of theories of the body and identities has provided new analytical tools to think about emerging health technologies, including reproductive technologies, genetic technologies and cell- and tissue-based technologies.

Technologies of the body

To begin, I want to reflect on the anthropological and historical accounts of biology and the life sciences. Regenerative medicine can be better understood situated within a historical account of contemporary life sciences. Historians of science and of medicine point to how biomedical understandings of the body and of biological processes were produced. For example, as we saw before in Chapter 2, Hannah Landecker's study of how cell-culturing techniques developed revealed how what biologists 'saw' and how they made sense of cellular life changed in association with new technologies for visualizing, preserving and culturing 'life' in the Petri dish and laboratory (Landecker 2007). These innovative technologies made new ways of seeing possible. The research questions being asked by biologists today rest on the 'discoveries' and problems of the past – these are fundamental questions about the life of organisms and the generative capacity of bodies.

Haraway's historical study of classical biology, animal sociology and sociobiology points to the ways in which social realities were mirrored in understandings of animal behaviour, such as male aggression and female receptivity. She showed how the normative theories of animal social organization were constructed through a lens of sexist thought and visualization. She suggests that more recent theories of the immune system are indicative of the emergence of postmodern understandings of the body. She says:

> Bodies are not born they are made ... Organisms are made; they are constructs of a world-changing kind. The constructions of an organism's boundaries, the job of the discourses of immunology, are particularly potent mediators of the experiences of sickness and death for industrial and post-industrial people.
>
> (Haraway 1995:208)

In her analysis, changing theories of the immune system produce a post-modern subjectivity – where other and self are not easily defined or separable. A model of the system as protecting the self from 'invading' organisms is reconfigured in these recent theories. In the 1970s, immunologist Niels Jerne proposed a network theory of the immune system that was *self-regulating*. In other words, it could produce antibodies to antigens that it produced itself. According to such a theory, the system was characterized by:

> the concatenation of internal recognitions and responses (which) would go on indefinitely in a series of mirrorings of sites on immunoglobulin molecules such that the immune system would always be in a state of dynamic internal responding. It would never be passive, 'at rest', awaiting an activating stimulus from a hostile outside. In a sense, there could be no *exterior* antigenic structure, no 'invader', that the immune system had not already 'seen' and mirrored internally. 'Self' and 'other' lose their rationalistic oppositional quality and become subtle plays of partially mirrored readings and responses. The notion of the *internal image* is key to the theory, and it entails the premise that every member of the immune system is capable of interacting with every other member.
>
> (Haraway 1995:218)

Moreover, she argues that, in Richard Dawkins' recent evolutionary theory, although the 'organism' and 'individual' do not disappear, 'they are fully denaturalized. That is they are ontologically contingent constructs from the point of view of the biologist, not just in the loose ravings of a cultural critic or feminist historian of science' (ibid.:220). Hence, the boundaries of the body as a separate biological entity appear weakened and undermined, or 'leaky' (Shildrick 1997).

In the figure of the cyborg, Haraway argues for a revisioning of feminist subjectivity that goes beyond dualistic and essentialistic thinking. 'The cyborg is a kind of disassembled and reassembled, postmodern collective and personal self. This is the self feminists must code' (ibid.:163). Increasingly, science and technology denaturalize bodies, not via a form of technological determinism but through processes of disassembly and reassembling, a re-arrangement of sociomaterial relations. Contemporary biotechnology politics, or biopolitics, is a politics of life itself (Rose 2007) that challenges naturalistic assumptions about bodies. Biology is not simply about nature or the materiality of life, but is inextricably cultural. Culture and nature, science and politics are intertwined and inseparable. For Haraway, we are all chimeras, hybrids or cyborgs, assemblages of science, nature and culture. At stake, then, is a reframing of the meanings of having or being 'a body'. Bodies are both socially and biologically produced.

The concept of 'regeneration' is not new, for it has been an enduring problem for biologists (Cooper 2003). Melinda Cooper describes epigenetic and preformationist theories of generation from the seventeenth century and how these early theories sought explanations about why the organism *Hydra* and some other simple life forms (e.g. flat worms) could regenerate themselves when dissected into parts. The dispute centred around whether generativity was a property only of germ cells. Epigenetic accounts saw higher animals as capable of self-organization of form from unstructured matter, whereas preformationists saw development as predetermined, 'encapsulated in the germ line' and passed on by generations (ibid.:4).[2] Hence, following epigenetic accounts, the self-organization of an organism and 'regeneration' of its parts were understood in terms of a self-regulating mechanism within the parts (tissue and cells) of the whole. Regeneration in this model does not depend on sexual reproduction, but asexual methods of replication and repair. The Hydra was thought to exemplify this, as it generated whole organisms from the dissected parts. Moreover, since the nineteenth century, studies of 'monstrosities', teratomas (cancer of the germ cell) and teratocarcinomas came to be understood as the unregulated or disorganized multiplication of life, the former, 'a kind of autogeneration, the reproduction of the self as a clone. But while the teratoma is a genetic clone of the egg cell, its disorganized growth in no way respects the rules of morphological self reproduction' (Cooper 2005:21). Rather, its effects are destructive of the body. Hence, Cooper argues that:

> What is exceptional about recent developments in stem cell research is the fact that such monstrous possibilities are being exploited *as a source of regenerative tissue* . . . At the same time stem cell research calls into question the difference between the regeneration and reproduction of the body, between regenerative and reproductive medicine. The process of tissue regeneration is reconceived as an act of permanent autogestation, an embryogenesis which can be reenacted throughout life.
>
> (Cooper 2005:21)

She elaborates, in a later paper:

> It is in the field of stem cell research that the 'new science of life' seems to have most successfully 'overcome' the experimental limits of the human life span. In the guise of the immortalized stem cell line, it has succeeded in culturing an experimental life-form that is capable of surviving long after the *potential person* [emphasis added] ever could have – a self-renewing source of cellular differentiation that isn't canalized

into the morphological limits of the human form. A cell line derived from one embryo can be divided up and proliferated in different growth cultures, without encountering predetermined limits to its growth; it can be distributed and circulate amongst an ever-widening network of researchers and clinicians while continually increasing in volume.

The proponents of regenerative medicine inscribe these experiments in stem cell culture within a generalized economy of 'auto-production'. Regenerative medicine, we are told, is about to bring about a new era in reconstructive surgery where the industrial model of prosthetics, which substituted a metal or plastic replacement for the failing or maimed organ, will be replaced by all kinds of self-replenishing tissues. The point, according to a recent introduction to tissue engineering, is no longer to supplement biological life with the spare parts of the industrial machine but to 'reawaken' the body's latent capacity for self-regeneration (Bell 2000 cited in Cooper). What is being articulated here is a vision of technology which is no longer mechanical or industrial but bio-regenerative – an intervention that works with the body's own generative processes, before their birth into determinate form, in order to resuscitate them otherwise. If we can continue to speak of 'invention' here, it needs to be reconceived as a process of *auto-reanimation*.

(Cooper 2006:10–11)

Hence, following this view, regenerative medicine draws on fundamental ideas of biological processes of generativity and animation, redirecting cells to behave in ways that restore function or performance (for example, brain repair). Technology, too, is reconfigured, and the process of invention here is reconceived, so that, as we saw in Chapter 3, the distinction between persons and things becomes blurred, with implications for patenting and ownership of the 'invention'. The regenerative body is a body that has the potential to regenerate itself, not through sexual reproduction but via this process of 'autogeneration' or 'auto-reanimation' that Cooper describes. The scientists' intervention is in order to stimulate, 'direct' or control this process, using (innovative) methods and techniques for cell harvesting, extraction, culturing, multiplication and replication. Embryonic SCs – SCs derived from the blastocyst[3] – exemplify in a particular way the intersection between biology and culture, for, as one scientist put it in this interview: 'embryonic stem cells represent really an artefact of culture because in the body embryonic stem cells don't actually exist' (SS7).

In other words, the derivation process *produces or creates* the SCs, a view also expressed by the eminent biologist Anne McLaren. The dispute about whether they are patentable turns crucially on the boundaries between 'nature and culture', that is, how an 'invention' (cultural) is defined and understood, and what is regarded as a 'natural' entity.[4] Of course, what is especially intriguing and controversial is that embryonic SCs (and cells obtained from an aborted fetus) are derived from '*potential persons*'. That is, they both precede the development of a person *and* have a life, which extends beyond them, for they are 'immortalized' (see Chapter 7). Moreover, as an entity, how far the blastocyst signifies a person or has the qualities of a 'thing' is contentious. In other words, embryonic SCs reveal the problems of dualistic thinking, of a moral ordering of the world into persons and things, subjects and objects, for the boundaries between them are exposed as permeable and hard to fix.

What an SC is and the meaning of 'stem-ness' have been especially difficult to fix or define, as I discussed in Chapter 2, and there were considerable uncertainties about whether cells can be redirected from one 'differentiation pathway' to a 'different lineage': 'Trans-differentiation is how people are collecting bone-marrow cells and inducing them to turn into, brain cells or skin cells' (SS1).

Another scientist, also researching brain repair, and in answer to a question as to whether it is like transplanting a kidney, said:

A: Well except on the individual cell level, so it sort of, I sort of think of it as a sort of, almost a sort of bagatelle theory. I don't know if this helps but you know, when you do damage, you knock out, say in the brain, you knock out certain populations of cells and so you can think of there being holes, and what the stem cells do is *they put cells back in the holes*. (SS5; emphasis added)

Therefore, at the cellular level, the functionality of cells, though only partially understood, is regarded as open to manipulation and redirection, and, rather than simply replacing damaged cells, the implanted cells actively repair or regenerate the tissue.

Multiplications

Bodies, then, have a materiality, a physicality and corporeality, but are in an important sense 'fabrications' (Shildrick 1997), that is, they are culturally produced. The 'biomedical body' is a product of changing ideas or *discourses* about anatomy and pathology and the effects of medical *practices*. The notion of the body as *multiple*, a composite that is produced or *enacted* through the practices of pathologists, doctors, patients, technicians and others, goes beyond the material body and knowledge of it as an epistemological project and becomes instead a study of praxis (Mol 2005), in other words, a study of how practices together comprise the body as an object. Anne-Marie Mol says this isn't to say they create it, but rather they produce multiple enactments of it, which, as she puts it, (mostly) 'hang together'. She contrasts this with the idea of the body as fragmented, or the epistemological relativism of multiple realities, arguing instead that work is done by these social actors to make the body (and its diseases) coherent. Therefore, in her study of atherosclerosis, there are many ways in which the disease is enacted in different settings, within and outside the hospital clinic. Similiarly, Mechteld-Hanna Derksen's study of a TE laboratory shows how the engineers' practices have normative effects as they design and engineer a tissue-engineered heart valve (Derksen 2008). They use normative ideas about bodies in their work that, she argues, do not adequately take account of the lived experiences of specific bodies, those of the people who might have heart valves implanted. The difficulty, then, is that 'solutions' to the engineering problem of designing a heart valve comprising human tissue might lead to a product that is functionally ineffective, because it is based on a different idea of the body from the one in which it is ultimately implanted. Mol describes how, in the hospital, treating patients for atherosclerosis, work is sometimes (although not always) carried out to resolve or overcome discrepancies between the different versions of 'atherosclerosis' encountered – for example, the results of the diagnostic tests, the clinical examination and interview of the patient. In the TE laboratory, however, without direct access to patients, Derksen argues that their 'lived experience' cannot be taken into account in the production of the heart valve. This raises again the questions I considered in Chapter 5 about how technologies configure bodies and the normative assumptions that shape the innovation process and outcomes.

At the same time, in a more literal sense, regenerative medicine reproduces bodies, or rather bodily tissues and cells, through the multiplication of them in the laboratory, the 'expansion' of cells seeded on to a scaffold or grown. In the example described in Chapter 1, the product Myskin entailed taking a biopsy in order to collect skin cells from a patient and the transportation of the cells to the laboratory, where they are cultured, multiplied in number and reproduced in the Petri dish. Once in sufficient quantity, the cells are placed on a polymer film for transportation back to the patient (one body) in the clinic, some weeks later. The scale of this is relatively small, in so far as the procedure is designed to treat a single patient. However, the multiplication of cells to create an 'off-the-shelf' product such as Apligraf, using cells from neo-natal foreskin, as we saw in Chapter 5, is larger in scale, designed to treat multiple patients (many bodies) and produce 'sheets' of skin – enough to cover a 'football field'. The transformation of bodily tissues into 'products' in the shape of small (or large) discs of skin is, as we've also seen, highly provocative and challenging to regulatory policy and institutions. Containment is difficult, both in terms of constructing regulatory categories but also because these cells become highly mobile, circulating widely within 'tissue economies' (Waldby and Mitchell 2006).

In another sense, bodies do become 'fragmented', dismembered or disassembled when parts are excized, tissues are dissected and cells are extracted. These processes of excision, dissection, extraction and manipulation can be read as a threat to bodily integrity only in so far as the body is conceptualized as having a form and boundaries that are distinctive and identifiable. However, when we see the materiality of the body, corporeality, as a production, or performance, something to be enacted in Mol's terms, then such threats to integrity appear to dissipate, not to have the same significance or be threats at all, for fears about the objectification and commodification of the body rest on a dualistic distinction, a humanistic concern, which is itself problematic, as it relies on a distinction between persons and things that is far from straightforward (Hoeyer 2007). Instead of seeing a body as fragmented by these practices, it is possible to argue that it is materialized in multiple ways *through* the processes of excision, dissection and so on. The meaning and significance of the materialization of multiple bodily 'fragments' are highly ambiguous and uncertain.

Self and other

These dilemmas and ontological problems are especially highlighted by the practices of abortion, fetal-tissue collection and transplantation. I will explore these more fully in the next chapter, but for now I want to draw attention to the ethical controversies surrounding these practices. First, the extent to which a fetus is viewed as a part of the woman's body; second, how personhood is attributed to a fetus *in utero*; and, third, how, once separated from the woman's body, it becomes available for use in research (or therapies).

As we saw in Chapters 2 and 5, fetal tissue has a range of potential uses, and its collection and use in embryology, vaccine production and medical research have a long history (Maynard-Moody 1995, Morgan and Michaels 1999, Morgan 2002, 2009, Kent 2008a, 2008b). This clinical scientist, working on brain repair and neurodegenerative disease, described both the use of fetal-tissue transplants using 'primary cells' and research to produce SC lines for brain repair:

A: The only cells that have been shown [. . .] in humans to actually work in terms of cell therapy are primary cells; cells taken directly from the fetus and put into the adult, without expansion.

Q: And is this from brain tissue?

A: So this is from fetal brain tissue yes. And very specifically they have to come from the specific area from which the cells that are lost in the adult are developing. So I mean obviously we work on Huntington's where the area of the brain that degenerates is the striatum, which is a particular nucleus within the brain, a particular set of nerve cells. So if we want to treat someone with cell replacement, we have to take the developing striatum from the fetus. So you can't take any other bit of brain, it specifically has to be that bit. And you can take those cells and put them into the area in the adult where the cells are being lost. And they will grow and develop and do the job. That's taking the cells with . . . we prepare them obviously, we have to prepare them for transplantation. But we're not culturing them, we're not expanding them, we're not extracting stem cells. So they're the primary transplants. (CL5)

She contrasted this 'replacement model', which uses cells 'grown up in the developing brain, they've got all the instructions to become the cell type you want. So they've been taught, they're committed to that particular cell type. So they work', with efforts directed at developing SC lines that need culturing in the laboratory but that had not, at that time, been successful or shown to have any clinical benefit. Importantly, if an SC line could be developed, because SCs are 'self-renewing', the need for fresh fetal tissue would be reduced. The technical challenges, therefore, were to produce a stable SC line that could produce 'well-characterized', healthy cells for implantation. That is, the properties of the cells could be controlled and quality managed. The stem cells could also be obtained initially from fetal tissue.

> Now you've got a window in a way because development is progressing slowly and for example the window we're looking at, in order to take cells for Huntington's disease, is between about eight and ten weeks gestation in a human. And during that time in the brain you will have still some stem cells, a mixture of progenitors and a mixture of differentiated cells. (CL5)

Another scientist, investigating the treatment of stroke, explained the difference between a simple 'replacement model' and regeneration like this, suggesting that two processes were interacting in the animal experiments they had conducted using fetal-derived cells (see also Chapter 2):

> And in the animal models we use, typically the animal might loose as much as 60 per cent of one whole hemisphere, so cortical and sub-cortical regions, so a huge chuck of tissue; billions of cells. Now what we do in those experiments, we in-graft in maybe a couple of hundred thousand stem cells right? So we're trying to replace billions of cells with a couple of hundred thousand, I mean we're putting in less than 1 per cent of the cells that have been lost. So the idea in any meaningful sense of replacing anything is absurd, so according to the classic model if you like, we shouldn't be able to see any functional improvement, but we do, and if we look at the functional improvement we see carefully and we follow it over time, what becomes increasingly obvious is that there's sort of two things going on and we can't quite disentangle them. [. . .] The ongoing neurodegeneration that follows stroke is halted, or at least reduced, and we seem to be inducing plasticity in what brain tissue remains. (SS5)

Since then, as mentioned above, a phase-1 trial using fetal-derived cells in human stroke patients has been approved and began in the United Kingdom in 2010.[5]

The fetus, once aborted, is separated from its potential human biography and, if donated, is transformed into human biological material for use in research (Pfeffer 2009b). It travels from the abortion clinic to the laboratory or research site, and its productivity is mobilized (Kent 2008a). Yet, as I've suggested already, there are conflicting and contradictory meanings attributed to the experience of abortion and towards fetuses in different contexts and settings. Within regulatory policy, abortion law and professional guidelines, a fetus was sometimes treated as part of the woman's body; after pregnancy termination, it could be regarded as like other bodily tissue, but, at other times, a fetus was regarded as 'special', deserving of respect (Pfeffer and Kent 2007, Kent 2008a). The Polkinghorne guidelines on the collection and use of fetal tissue for research and recommendations by the Royal College of Nursing for 'sensitive disposal' drew on the idea of a fetus as a cadaver or corpse (Polkinghorne 1989, Royal College of Nursing 2002). As such, the fetus acquired attributes of a dead *person*, and proposals that there should be special arrangements for the burial or cremation of the miscarried and aborted fetuses reiterated such ideas (Pfeffer and Kent 2007, Kent 2008a). In the abortion clinic, we learned that some women thought of, and talked about, the fetus as a baby, but staff were concerned to avoid such a term, in order to minimize the potential distress for the woman undergoing termination (Kent 2008a). This research nurse, in describing her experiences of asking women who were having an abortion if they were willing to donate the fetus, said:

> They do ask – I have been asked – will the baby, once it's born, is it dead? You know. And I've also been asked, 'So you wanna take its brain out and use it?' You know, quite explicit terminology like that, which I think, you know, if I said to them at the beginning of the discussion, 'We want to take your fetus's brain out and use it for research', that's a really quite hard thing to ask, but if they ask me in that terminology, I will say yes, because there's no point in pussy footing around, you know. At the end of the day, that's what we are going to do. (CL6, research nurse)

In commercial settings, considerable discursive effort is made to separate the cells derived from an aborted fetus from its source, and, at the time of my research, 'adult stem cells' as a biological category was a term used to include cells derived from a fetus (Kent 2007, 2008b). However, for many outside the scientific community, it was understood as referring to cells derived from an adult person. Of course, the ways in which scientists, nurses, women, regulators and others talk about materials obtained from pregnant women are laden with a range of moral values and concerns. Medical concern for the fetus *in utero* is manifest in therapeutic practices that construct it as a 'patient' or person (Casper 1994,1998, Williams *et al.* 2001, Williams 2005). However, following pregnancy termination, fetal remains can be regarded as 'waste' or 'just tissue' (Morgan 2002), and women who terminate a pregnancy can be seen as especially wasteful (Ariss 2003, Kent 2008a,). The fetus, therefore, is materialized as a boundary object (Bowker and Starr 2000, Hoeyer unpublished) in multiple ways and has a highly ambiguous and indeterminate form, mirrored in the multiplicity of meanings attributed to it.

In the United Kingdom, abortion law does not rest on a woman's right to choose what she may do with her own body (see Chapter 5).[6] She needs the approval of two medical practitioners before an abortion can take place, and then only if the criteria set out in the 1967 Abortion Act[7] are met. The law is premised on public-health concerns to protect the woman's health.[8] Pregnant women seeking termination are framed as in need of protection from themselves and from the coercion of others. The procedures for seeking consent to undergo an abortion and also to allow the aborted fetus to be used for research can be seen as paternalistic (Pfeffer and Kent 2007) and undermining women's autonomy.[9] As we saw above, these women's sexual history is also regarded as making them potentially more risky donors, who are more likely to produce biological material that is contaminated, unsafe and unsuitable for research or therapeutic use (Kent 2008a).

Rachel Ariss's (2003) analysis of how the reproductive labour of women is tied to notions of wastefulness is especially useful in thinking through the relationship between women and their fetus in the abortion context (Kent 2008a). Following Kristeva, she argues that the act of abortion produces 'abject waste', that, 'if abortion is seen as abjection, it is one route through which fetal tissue can come to be seen as waste' (Ariss 2003:270). For her, harvesting tissue from a dead fetus[10] undermines the subjectivity of the woman and assertion of her bodily boundaries by creating 'the possibility of future out-of-her-control bodily connections to the world' (ibid.:271). From her discussion, we see that donation of 'waste' fetal tissue is sometimes seen as redemptive and the act of 'good citizens'. Ariss seems to want to protect a reformed humanist subjectivity, to reclaim the bodily boundaries of women, and with it political self-determination.

In summary, therefore, as is well recognized, pregnancy and abortion are sites where the ontological problem of the relationship between 'self' and 'other' becomes evident (Kent 2000).[11] The body of the pregnant woman is connected to that of an 'other' – the genetically distinct but materially connected fetus/baby. Maternal and fetal circulations mix; the bodies of the woman and fetus are 'entangled'. And women who have been pregnant are, in another sense, 'chimeric' bodies in which the fetal cells circulate (see Chapter 7). The law is a blunt instrument that attempts, unsatisfactorily, to fix and stabilize these bodies. In Britain, the law attributes rights to the woman to do what she chooses with her body, although these rights are conditional, and clearly there are limits set where the decision to terminate a pregnancy has to be approved by medical practitioners (Sheldon 1997). Medical practices that intervene in the woman's body during pregnancy in order to treat the fetus construct it both as a 'work object' and as a 'patient' (Casper 1998, Williams 2005). Moreover, it has often been suggested that such interventions are not always in the best interests of the woman, whose body is commonly constructed by such practices as a 'container' for the fetus (Oakley 1986, Petchesky 1987, Bordo 1995). If aborted fetal tissue, or cells derived from the fetus, is transplanted into another body, additional questions about the relationship between 'self' and 'other' come into view and might raise identity issues, as I shall discuss below.

We investigated the views of women about fetal-tissue donation for medical research using focus groups. Six focus groups of women of reproductive age were held. Two of the six focus groups were with those who had not had an abortion (FG1, 2), a total of ten women. Four groups were with women who had experienced an abortion (FG3, 4, 5, 6), and, of these, three out of the thirty-one women had been asked to donate fetal tissue for research, and two had agreed.[12] A number of themes emerged in relation to the meaning of 'life' and what it means to be 'human'. The groups engaged with problems of how to think about a fetus, its ambiguous status and the different meanings attributed to it. Those who had had abortions described their experiences and feelings about what had happened to them and how it affected their views on tissue donation.

Being human, donating tissue for research and therapies

Early on in discussions, the first group (who had not had abortions themselves, but knew women who had) wrestled with ideas about what it meant to be 'alive' and uncertainty about whether an aborted fetus might be 'kept alive' if it were to be used for research:[13]

FG1F2: Would it be alive whilst it was being experimented on?

MODERATOR: What would your thoughts about that be?

FG1F2: I don't know, I mean, I'd probably ask about that. But do you know?

FG1F4: I would have assumed that to have an abortion your fetus is killed anyway, that would be my assumption but, but it would change things I think, it would change things for me if it was alive as such.

FG1F2: Because presumably they can be kept alive for a bit?

FG1F5: I think ... I don't think that they would be, I mean, there would be a lot of sort of ethical and legal implications of experimenting on them.

FG1F2: Would there be any point in experimenting on them, though, if they weren't?

FG1F4: What do you mean by 'alive'?

FG1F5: Yeah, that's ... I suppose there's varying degrees of it, isn't there.

FG1F4: It's quite important though, isn't it?

FG1F2: What does life mean? Having a beating heartbeat I suppose.

FG1F5: Do you mean whether it would be capable of being kept alive after it was ...?

FG1F2: Yeah, would it be kept alive in order to be experimented on?

Like the women described above by the research nurse, concerns that a fetus was dead were important, because this was linked to ideas about whether it might feel pain. And whether or not 'being dead' affected its usefulness for research was considered as significant, and yet what it meant to be dead or alive was far from clear (see Chapter 7).

The extent to which a fetus might be regarded as a child, a baby or 'something human' was explored and considered an influence on whether or not a woman might want to donate it for research:

FG1F4: I suppose it's about the way you view your fetus. Not everyone who has an abortion would look at it that it's something that they're willing to dispose of, and you might view it as something that has somehow got some sort of spirituality about it, and you'd want to bury it, and, you know, you'd see it as a sort of human being, I suppose.

FG1F5: As your child.

Women in all focus groups switched between the term baby and fetus, and the former was more likely to be used to refer to a 'wanted' pregnancy, and fetus used for an 'unwanted' one (Pfeffer 2008). Underlying discussions were considerations about what it meant to be human, and what, if anything, distinguished a fetus from other body parts, such as a kidney or liver. Was it different, and if so in what way?

FG1F1: I'm not sure. I'm just wondering why we're making these distinctions between the kidney and the fetus, or the liver and the fetus. What's the difference?

A baby was something that 'had a face', 'had a heartbeat' or could 'feel pain', or even was, in some sense, 'conscious'. In certain circumstances, it was a potential 'person'. The developmental age of the fetus could also influence their feelings about whether or not it might be donated for research. In early pregnancy, it could be viewed as 'a bundle of cells' and therefore less than human. One woman, who had been pregnant but had not had an abortion, said:

FG1F1: Having been pregnant, I was just thinking back to early stages of pregnancy, and I didn't – I know some women do as soon as they know they're pregnant – have an attachment to what they feel is growing inside them, but it took me a while. It wasn't really until I felt the movements that I started to feel that there was someone there. So although I knew quite early on that I was pregnant, in the first few weeks it was just like a bundle of cells to me.

And, in another group:

FG3F7: I do think it does depend on what stage you're at and some women do see it as a baby straight away, others see it as a bunch of cells.

A scientist I interviewed also emphasized what he saw as an important distinction between a fetus and the cells derived from it:

A: The issue for us is, 'When does human fetal material stay as human fetal material, and when does it become a tube of DNA or RNA?' And so, yeah, we have some collaborations, but the fetal material handling takes place here [. . .] and it's – the parallels for stem cell research, or embryonic stem cell research, are – kind of like saying, 'These are cells, they're not embryos any more, they are cells', and I think that's really important, and that's one that again is missed in terms of portrayal. And so, we receive fetal material. . . . When I put some hepatocytes, liver cells, into culture, they are cells; they are no longer a fetus. *They're human cells but they are no longer human in terms of how we understand human life*; they could as well be mouse or rat in terms of they are hepatocytes; they are cells, primary cells. (SS2; emphasis added)

Although not a fetus, they are human cells, but their human origins are underplayed here, as they 'could as well be mouse or rat'. For him, once extracted from the fetus, they are 'just cells'.

Having children seems to influence how some women think about abortion, and, in our groups, some, since having children, had changed their views about previous abortions, while others saw their earlier decision as in the interests of their existing children (Pfeffer 2008). As Naomi Pfeffer suggests in her analysis of our data, the thought that the fetus might 'live on' if it were used for research provoked a sense of continuing responsibility for it among the women, which detracted from their support for donation. The extent to which they viewed a fetus as distinct from, or similar to, organ donation was also a strong influence on their attitudes towards donation.

The aim of an abortion is the physical removal of the fetus from within the woman's body. And common law regards the fetus *ex utero* as no different to say a gall bladder or an appendix. Participants offered three different reasons why a fetus aborted on social grounds is unlike other excised body parts. First, women initiate an abortion on social grounds – it is their decision whereas most other medical procedures are recommended by a doctor. Second, and related to the first, the purpose of abortion is to separate irrevocably woman and fetus. However, while physical separation might be complete, some participants said an attachment of sorts had persisted . . . [but . . .] 'I'm not attached to my kidneys really' (FG5F8). . . . Third, the fetus is both unique to women and yet, unlike her other body parts, not identical to her because it is a product of sexual intercourse and partly 'made' by its biological father. Put another way, 'it's not yourself'. (FG3F3)

(Pfeffer 2008:2548)

Despite recognizing that 'it's not yourself', this same woman (FG3F3) worried about 'her DNA being out there', provoking concern from others that, once donated, the tissue was beyond your control and could lead to cloning, and 'You open the door, and you're looking at a duplicate of yourself' (FG3F2). This didn't acknowledge the distinct genetic identity of the fetus. Or that, 'you give your fetus away and then in years to come the government will be like, oh hello we've got a baby for you' (FG3F7). Ownership of, or responsibility for, the fetus was therefore seen as potentially extending beyond donation.

FG3F3: I would worry about my DNA being out there as well.

FG3F7: Yeah, that's a point, cause actually, obviously I don't know a lot about it, but I understand that it is the cells and tissues which can be used to grow all the things that make up you, and so that's the point . . .

Some women expressed what Pfeffer calls a 'duty of care' for the fetus, which extended beyond the abortion itself. Others were less concerned:

FG3F7: Had I been asked, I would have definitely said yes, because I'd made the choice to have the termination, and therefore I'd detached myself from carrying this, from developing it into a baby, so I'd made the choice, I'd detached myself, so if it can be good for somebody else, me personally, I'd say go ahead.

MODERATOR: What do other people think about that?

FG3F3: I think if I wouldn't be able to make that decision, I think, I think it would be in the back of my mind that there was some ongoing process, whereas I really want, personally, I'd just want it to be over and then done, and then not something that I would think about again.

The idea that there was 'some ongoing process', or that the fetus lived on, presented a real dilemma for them.

Identity was linked to the body but not to individual organs. Although some participants were strongly supportive of the idea of donating their organs (and a significant proportion of them were registered organ donors or blood donors),[14] the idea of donating one's body for medical research and education was, in this example (contradictorily), a problem:

FG1F4: But even in that case, I mean, it depends, I mean, you know, depending on how you, what you thought about organ donation more generally, or just donation of body parts, or, you know, after death. Even if you did, you know, your decision could be swayed by what you thought about that.

FG1F5: Yeah, I'm sure you're right. I just imagine it being a lot more, a much more sensitive issue. And even people . . . I mean, I've got funny, like, irrational views, and I'm a sort of philosopher, like, I sort of, I'm very happy to give my organs away after death, but I don't want, I wouldn't want to offer my body for medical research as sort of to be a medical student's body – it just gives me the creeps, even though I don't see my body as being, you know . . . I just don't like the thought of it, yeah. And that's irrational, and that's purely emotional, but I would very much want to give my organs away after I die.

FG1F3: But only if they went to save other people? What about if they got cut up or stuff?

FG1F5: Yeah, I may know.

FG1F5: You're not attached to your organs – 'Oh, look at my lovely kidneys!'

FG1F1: Yes, exactly. I think it's also because I've known medical students who probably sort of take the mickey out of it!

FG1F3: Oh right!

FG1F1: Yeah, that idea of sort of lying on a slab and not being buried for years. Yeah, I think if it was being removed either for donation or research, then my body could be buried and my family could have a funeral, and, you know, that letting go, you know, that would be quite important, so I suppose there is a bit of rationality.

FG1F2: It's your identity, isn't it, in the way your kidneys or whatever aren't your identity. You never see them, I suppose. But you like to think that your identity remains in the way that you left it.

They didn't feel the same way about their kidneys or other organs as they did their body. However, in comparing donating a fetus and their own body for science, or donating organs for transplantation, different ways of thinking about fetal-tissue donation came into view. It could be like donating an organ, which might be okay, but donating your body, not being buried, wasn't as easy to accept, because somehow having your body dissected by medical students was more challenging and disruptive of bodily integrity and of one's identity.

The idea of *transplanting* fetal tissue into another body seemed at first to be repugnant to some participants, who immediately expressed concern about it. In particular, distinguishing between *cells*, which might be transplanted, and an *arm* or something that you could see and identify as a body part became important:

MODERATOR: I want to ask you a few more 'What if?' kind of questions. So, what would you think if the cells were being transplanted into another person, stem cells from a fetus?

FG1F5: I would want to know why. What purpose is it going to achieve? Because the instinct, as soon as you said that, was 'Urgh', so I would want to know what the purpose was and why.

FG1F3: But don't you think the purpose would obviously be because it was going to help them in some way, so just assume that?

FG1F4: But I'd want to know, not that it's going to help with their new kind of botox or something, but is it actually for their health? It's still tricky to know.

FG1F5: I think, it was just that my personal reaction was 'Urgh', so I think that if I had known that that was going to happen, I would definitely want to know what the purpose of somebody being injected with stem cells was going to be, and how that was going to help them.

MODERATOR: You're sort of going 'Um'. Can you say it out loud?

FG1F3: I don't know, I just, just to me, I don't know. It's . . . At the moment, I just can't physically or mentally work out what the stem cells are doing in somebody else's body, if it is . . .

FG1F2: They could grow a new lung or kidney.

FG1F4: Or a head.

FG1F5: They've got regenerative properties, stem cells, and they do it at the moment. I don't know if it is actually proven. And they say that we'll sort of inject it into a cancer, the idea being that we'll regenerate it.

FG1F3: What sort of flashed into my mind, when you said that it's going into somebody's body, was just like the arm of a fetus or something. Like the mouse ear. That's what I was thinking of. I would definitely want to be told that these are going to be tiny cells that you can't even see, or whatever. [laughter]

The imagined possibility of recognizable body parts being transplanted was deeply problematic, but if it involved 'tiny cells', lacking in specific form, it would be more acceptable.

The connections between identities, being human, the body, body parts and bodily tissues or cells were far from straightforward in the focus groups' accounts. They had to be made, and easily became entangled and complicated, difficult to define or fix. An aborted fetus was human, a baby, cells, waste, research material, alive and dead, deserving of respect but also something that could be taken apart and used, transplanted into the body of another. It had multiple meanings and significance; its status was uncertain and controversial, as could be expected.[15] In discussing donating fetal tissue for research and therapies, the women drew parallels with organ donation, but also identified differences. The fetus was, at the same time, a part of them, but not them.

Such views can usefully be contrasted with attitudes towards the donation of other reproductive tissues – ova, umbilical-cord blood, placenta and embryos, as I shall discuss in the next chapter.

Plastic bodies

There is a range of different sources of tissue used in regenerative medicine, as we have seen already, including embryos, ova, fetuses, umbilical-cord blood, bone marrow, skin and cartilage. In a booklet produced by the UK MRC to inform the public about its funding of SC research, it explained:

> Stem cells are immature cells that can divide to produce more of themselves (self-renewal) or give rise to the specialized cells that make up our organs and tissues (differentiation). They are found in many places in the body and are present from just after fertilisation of an egg right through to adulthood. Unlike specialized cells, different types of stem cells can generate many different types of cell, such as the beating cells of the heart or the insulin-producing cells of the pancreas. Scientists all over the world are studying stem cells to learn what makes them different from specialized cell types. They are also trying to understand what causes stem cells to renew themselves and what causes them to specialize into other types of cell. As the investigators learn more about these processes, they hope that it will become possible to collect stem cells and direct them to form particular types of cell from which to grow supplies of healthy tissue. This tissue could then be transplanted into people who are ill or injured to replace their own damaged cells and make them well again.
>
> (Medical Research Council 2008)

Stem cells are, in biological discourse, categorized according to their source, different properties and potentialities – usually described in terms of their 'potency'. Embryonic SCs are *pluripotent* and have the ability to form all of the different cell types that make up human organs and tissues, as well as sperm and eggs.[16] SCs that have already partially specialized into a particular broad cell type are *multipotent*, because they can still differentiate into different types within that (e.g. blood-forming SCs found in bone marrow can specialize into red or white blood cells or platelets, but cannot generate brain cells). Other, partially specialized cells can only go on to form a single type of cell and are *unipotent* (e.g. SCs in the testis that go on to produce sperm cells) (Medical Research Council 2008). However, most recently, with the success of teams in Japan and America described earlier, it was discovered that cells previously considered specialized – for example, a skin cell – could become pluripotent, and fibroblasts obtained from a fetus were used in one of these early experiments to produced 'induced pluripotent stem cells' (Takahashi *et al.* 2007, Yu *et al.* 2007) (see Chapter 2).

The ability of scientists to control the ways in which the cells behave, their functionality and characteristics rests on notions of *plasticity*. Rather than the properties of cells being fixed, they appear to be manipulated, open to direction and control through the intervention of the scientist (Landecker 2005). Technical reformulation and reordering of the body and its components become possible at the cellular level. Body modification, then, can be reconceived, not simply as reshaping the body using cosmetic surgery or body-building, enhancement technologies, piercing or painting, but at the fundamental level of the cellular 'building blocks' (Davis 1995, Andrews and Benvenisty 2005, Hogle 2005). Cellular plasticity refers to the generative potential of *all* of the cells of the body and, in so doing, goes beyond the preformationist ideas of earlier biological theory. Therefore, through these practices and techniques, bodies become 'denaturalized', and new objects of science, new categories and new entities are produced. Associated with this 'leakiness', the productivity of the body is mobilized, 'life' is multiplied, and the possibility of 're-engineering' the body comes into view.

Since the 1990s, TE can be viewed as aiming to overcome the shortcomings of earlier biomedical technologies, where implanted devices failed or became worn out, and organ transplants provoked rejection by the host body (Langer and Vacanti 1995, Williams 1997). The limitations of these earlier implant technologies relate to the properties of the biomaterials used in them and the design features, which create stress and wear through use. Organ transplantation, now a commonplace procedure, depends on the continuing use of immunosuppressive drugs to prevent tissue rejection. TE can be seen as drawing on 'concepts of animation, bodily (re)generation and transformability' arising from its disciplinary background and conceptualization of morphology and 'topographical space' and time (Cooper 2008:105). Cooper brings into view an intriguing perspective on how early ideas about bodily time and movement informed the development of these biomedical techniques of prosthetics and implantation. Based on these ideas, she says, 'the whole aim of organ transplantation is to suspend animation to then revive it elsewhere', and that 'organ transplantation like prosthetics is an art of "spare parts"' (ibid.: 110). Both practices are based on a mechanistic vision of animation and transplanting an organ or implanting a device that has a *morphology* that can be integrated into the recipient body. 'As modes of biomedical intervention, prosthetics and organ transplantation need to arrest the process of morphogenesis to then work with the frozen morphological form' (ibid.:111). In contrast, TE and regenerative medicine, rather than simply reconstructing the body, seeks to *generate* the form, drawing on the principles of 'organogenesis'. As such, its aim is to intervene in the dynamic process of growth and generation of tissue, rather than simply replace it. The seeding of cells on to scaffolds and culturing and expansion of cells in a bioreactor are aimed at directing the cells into a particular form and function.

Regenerative medicine works through the continuous variation of force fields, and it is from this level up that it attempts to determine the emergence of particular tissue qualities (density, compressibility, elasticity), properties, and forms (cell morphology, differentiation, organ morphology and structure).

(ibid.:112)

In other words, regenerative medicine aims to re-assemble the body. In so doing, bodily time is seen as a perpetual process,[17] rather than as needing to be suspended and reanimated, as in organ transplantation – 'far from requiring the solidification of the organ (its biological inertia) the technicity of the intervention works with and exploits the active responsiveness of living tissue' (ibid.:113). This is an issue to which I will return in the next chapter.

So, what are the implications of thinking about TE in these terms for understandings of the self and subjectivity? And in what sense does regenerative medicine reproduce the self, or produce split subjectivity or mobile subjectivities? It has been suggested that bodily integrity and subjectivity are undermined by such developments, that modernist conceptions of the subject and personhood are irrevocably damaged or irretrievably lost. Yet I want to consider more carefully whether the source of cells and their types produce different effects for how we think about these issues.

Culturing cells and regulating the self

In an earlier paper, we suggested that, 'it is important empirically to disentangle different types of tissue-based technologies because they carry different meanings and produce different sets of relations' (Kent *et al.* 2006a:3). Focusing on the production and delivery of ACI (described above in Chapter 5), we argued that the distinction between autologous and allogeneic applications of TE was significant within policy debates about appropriate regulation and governance arrangements (during the period leading up to the drafting of the RATMP) (Kent *et al.* 2006a). Autologous applications, where the patient's own cells are extracted, manipulated and reimplanted in their body, were often regarded by scientists and policymakers as relatively low risk and ethically uncontroversial. Most importantly, from a public-health perspective, the risks of contamination or viral transmission between patients was perceived as low compared with allogeneic products, where cells might be widely distributed between multiple recipients. This isn't to suggest that there were no risks associated with autologous cells, nor that risk-management practices weren't important –there was clear evidence of the need to ensure that the 'chain of custody' for the cell sample was carefully managed (Lynch 2002), and that techniques in the laboratory ensured the quality of the cells returned to the patient. Nevertheless, as one scientist put it in an interview:

Well I think the ethical issues are immensely important here. And I used the phrase that autologous tissue engineering is an ethics-free zone. There is an ethical issue there if you are taking the patient's own cells and using those cells to generate the patient's own tissue the only time ethics would really come in is if you're using gene transfer in that process. But even then I don't think that's a big issue. If you're using allogeneic sources, you have to, obviously there are scientific issues and they, you have to ask the question, then where are the cells coming from, whose intellectual property is it, whose cells are they? Did the donor know that they were actually donating those cells? And the answer is, in most cases, no – they had no idea. And I think that those are issues which have to be very carefully answered. (A-EU6)

According to this view, therefore, ethical questions arise primarily in relation to *donated* tissues and cells and the *transfer* of them between persons – what has been referred to as *intercorporeality* (Waldby 2002b). We suggested that ACI exemplified a type of 'extracorporeality', 'which represents an externalized and extended vision of the self and materiality of the body' (Kent *et al.* 2006a:17), and we suggested that 'discussions of autologous products such as ACI mobilize a discourse of individualism and selfhood that obscures and resists the potential destabilizing effects of the technology on the self' (ibid.:17). What seemed to be underlying policy debate and at least some scientific accounts was a reiteration of individualism and humanist bioethics, which asserted a naturalistic view of the body and conceived of the process of extraction, amplification and reinsertion into the body as undermining neither bodily integrity nor liberal notions of subjectivity. Moreover, in this analysis, autologous applications could be seen as selfish (or self-serving) in attempting to remake the self and enable self-repair. In addition, the ability of the body to recognize 'itself', that is, cells derived from it, was cited as an advantage, avoiding the immunological problems associated with donated tissues from an 'other'.

In contrast, donating tissue and cells, which is frequently understood as an act of solidarity through the 'gifting' of them for use by others, implied a process of tissue exchange and, following manipulation and processing, 'biovaluable amplification' (Waldby 2002a). We saw this as more like the literal kind of intercorporeality to which Catherine Waldby referred and which was associated with ethical concerns about consent, ownership (intellectual property), risk and benefit. The higher risks of allogeneic applications were often understood by our interviewees in terms of potential transmission of disease, rejection, batch control and quality management (Kent *et al.* 2006a) and, where cells were cultured in animal-derived media, interspecies mixing (Brown *et al.* 2006a). Hence, in contrasting autologous and allogeniec applications of TE, we can point to important ways in which they are distinct from each other – especially in terms of the source of cells and the character of the exchange relationship (between 'donor' and recipient) and, from a technical point of view, the distribution and management of risks. However, the extent to which these different forms of TE have differing implications for a liberalist view of human subjectivity is worth revisiting.

First, it has been suggested that our analysis underplays the *engineering* of the technology and the degree of manipulation of autologous cell applications, and the engineering problems associated with this (Derksen 2008),[18] although, as we indicated, the ways in which 'engineering' of human tissue could be conceptualized were themselves a focus of debate in formulating the regulations for what later became known as 'advanced therapy medicinal products'.[19] I recognize that autologous products are *fabricated* – as the extraction, transportation and processing of them are elements of a production (labour) process.[20] Work is done to intervene in the body, culture and grow the cells, seed them on to a manufactured scaffold and surgically reimplant them. This is not a simple form of self-healing but, as Derksen suggests, requires a type of 'interaction' with researchers/manufacturers/engineers. Moreover, as in the case of the heart valves she describes and the cartilage and skin products I have described, the protocols developed for these processes are subject to standardization. Similarly, allogeneic products are extracted and travel between donor and recipient bodies via the laboratory or 'cell factory'. Subsequent iterations of the regulatory framework require similar standards and safety controls for both types of product.

Second, it might appear that our analysis did not question a naturalistic view of the body, that, in suggesting that autologous applications of the technology 'remake' the self, we failed to question the notion of 'self' and its relationship to the body and others. Instead, although regeneration of the body can be understood as a form of 'autogeneration', far from seeing this as reinstating unproblematically the naturalistic body, the notion of 'extracorporeality' points to the permeability and instability of bodily boundaries and the relationship of the body to others. Interestingly, Cooper's account of tissue regeneration does not distinguish at all between autologous or allogeneic applications, but she does say, 'regenerative medicine is not so much interested in the reproduction of the human person as in the stem cell's capacity for multiple differentiation and indefinite renewal, in excess of the limits of the developmental human morphology' (Cooper 2008:147). In disputes about the personhood of the human embryo and (allogeneic) cells obtained from it, biologists have indeed emphasized the difference between reproducing a human person and human tissue, but, in relation to autologous applications, this distinction has been obscured, and it is precisely for this reason that, within Europe at least, autologous applications received wider political support. They were ethically less problematic or 'risky' and, at least in some accounts, technically easier to control and manage.

Third, our use of the term 'intercorporeality' emphasized a literal reading of exchange relations, as described by Waldby (2002b). However, as has been pointed out to me, the work that Waldby draws on, by Weiss (1999), sees embodiment as always in relation to others. Feminist readings of embodiment emphasize this, and, following Weiss, bodily integrity can be understood in terms of continuous change and flux. In these terms, from a phenomenological perspective, patients can experience tissue-engineered products (autologous or allogeneic) without disruption to either bodily integrity or identity, but this is an achievement rather than 'natural'. A 'good embodied life' may mean different things to different people and is an accomplishment whereby, for example, the prosthetic device becomes relatively transparent in the life of that person, and bodily change is experienced as a positive aspect of everyday life (Derksen 2008). Currently, there is a lack of research on how people *experience* regenerative medicine or tissue-engineered products, in part because, with so few products actually having reached the clinic, there is still a relatively small population to study.

Beyond limits: materiality and subjectivity

Other investigations of embodiment and identity that draw on phenomeno-logical approaches highlight how biomedical interventions, specifically surgical cuts of the corporeal body, can have psychic consequences. It has been suggested by Margrit Shildrick that dominant rhetoric around organ donation as a 'gift of life' underplays the psychological impact on recipients of receiving a transplanted organ. As a result, while, on the one hand, recip-ients are encouraged to think about the donated organ in mechanistic terms – a replacement heart as a replacement pump – at the same time they are encouraged to write letters of thanks to donors' families and recognize the provenance of the donated organ (Shildrick 2008a, Shildrick *et al.* 2009). Other anthropological studies too have explored these new forms of 'biosociality', how, in the organ, recipients perceive themselves as having a kinship relationship to the deceased donor (see Chapter 1) (Fox 1992). The self is understood as *embodied*, but, rather than simply a container for the self, encompassed by the outer layer of skin that defines bodily boundaries, bodily boundaries are 'leaky' (Shildrick 1997). A phenomenological approach sees body and identity as inextricably linked: 'the materiality of flesh and blood is inseparable from the question Who am I?' and is 'the very condition of being a self at all', so that, 'when corporeal changes occur, they inevitably signal changes to the embodied self' (Shildrick *et al.* 2009:36). In Shildrick's analysis, changes can be externally visible, morphological changes (cosmetic surgery, amputation, separation of twins) or internal ('solid' organ transplants); both can have effects on identity. However, whether there are important differences between these bodily changes isn't clear. What is notable is that what con-stitutes bodily change here is morphological. She argues therefore that:

The image of the skin as some kind of inviolable boundary no longer makes sense, for it is continually breached by a series of biomedical technologies in the interests of maintaining the illusion of a normative body . . . As Claudia Benthien . . . outlines, [historically] the porosity of the skin, far from being a matter of disgust or even horror, was an essential component of good health. The skin was seen not as a protective layer against the insults of the external world – as a barrier that might hold disease at bay – but as the conduit through which the materiality of internal disturbance might leave the body through bleeding, scalding, scarifaction and other 'therapeutic' techniques which promoted leakage from the body [. . .] a quite literal and active leakiness was seen as highly therapeutic.

(Shildrick 2008a:33)

In her analysis, contemporary biomedicine makes efforts to reseal the body to fix and secure bodily boundaries, making it good and restoring it to the appearance of a normal (ideal) body. This effort, she suggests, is deeply problematic in the case of organ transplantation because, although the skin might be stitched together, there is a disruption to embodiment and identity that, in her view, should not be denied. In contrast to Derksen's view that change and multiplicity of experience are features of embodiment, Shildrick sees disruption to the self as inevitable and as producing negative effects in some patients. Therefore, in proposing that organ recipients be encouraged to acknowledge the disruption to embodiment that can follow transplantation, she suggests they might then be able to come to terms with 'bodily hybridity'. Her proposal for resolving the 'troubling' effects of heart transplantation is to 'embrace concorporeality: the point at which bodies cross over into one another' (Shildrick 2008b:43). By recognizing that the phenomenological self is intercorporeal, she extends that idea to *concorporeality*. This, she argues, is commensurate with a feminist bioethics.

My interest, then, is to consider whether this notion of concorporeality might be helpful for thinking about regenerative medicine, and to note the distinctions between the entanglement of bodies through organ transplantation and the regenerated body, for, if we see the former as an example of the biotechnical art of 'spare parts' and the latter as qualitatively distinct, it follows that the implications for understandings of self vary. Concorporeality draws our attention to the mixing and co-residence of two bodies (parts), and the view of a transplanted heart as a fragment carrying with it attributions of the donor can be contrasted with the mixing of human cells grown in a laboratory and subsequently implanted, or the injection of suspended cells that travel around the body. The idea of 'hosting' the organ, which Shildrick suggests, doesn't quite adequately describe the diffuse intermixing of cells that occurs in regenerating bodies. (Such intermixing, of course, also occurs in transplantation; hence the need for immunosuppression.[21]) In her formulation, the form and shape of the reconstructed body are central, and yet it is precisely a shift away from morphology to organogenesis that bodily regeneration implies. Moreover as we saw in relation to the product Epicel, one of the earliest skin applications, which was cultured on animal feeder cells, the human–animal boundary and interspecies mixing have also been features and of ethical concern relating to tissue-engineered products and SC therapies (Brown *et al.* 2006a).

Let us also consider comparisons between cerebral implants, which deliver deep brain stimulation (DBS), fetal-tissue transplants and SC therapies in PD. The effects of cerebral implants providing DBS have been seen as disruptive of a person's identity, in spite of functional improvements.

> Unlike other prostheses [. . .] or organs, this new interface improves the patient's motor capacities, but also modifies other parameters which constitute the patient's identity, such as behaviour, mood or cognitive functions. DBS does not cure Parkinson's disease, but it can help manage some of its symptoms and subsequently improve the patient's daily activities.
>
> (Gisquet 2008:1847)

Despite therapeutic benefits, this study in France suggested that patients' experience of loss of control over managing their illness and their lives, and the consequences for identity, were attributable to effects of the implant and were neglected by medical practitioners. Managing the device detracted from a patient's ability to 'live normally' or live as he/she had before treatment began, and adjusting to living with the implant proved difficult. In Sweden, other patients (neurological and diabetic) preferred the thought of animal-tissue transplants to such mechanical devices, despite concerns that such transplants could represent a threat to identity (Lundin and Widner 2000). Attitudes towards xenotransplantation have been described as 'a moral quagmire', and, despite negative views of it overall, such views coexisted with the idea that, in specific circumstances, it was acceptable (Lundin 2002). The Swedish study, which compared the experiences of those who had pig cells transplanted into their kidneys to treat diabetes and of those who had fetal-tissue transplants to treat PD, provides some interesting insights into patients' views.[22] The location of the pig cells in the body, once implanted, was viewed with uncertainty, whereas the fetal cells were readily identified as located in the brain at the site where they were implanted.[23] As Susanne Lundin reported, one respondent said of the pig cells, 'they are, of course, so diffuse and small that one can't know where they are, or what they are doing with us', but was reassured by the doctor present that they would have been excreted, and 'if not we certainly don't transplant souls in this hospital'. Another patient, who had been transplanted with fetal cells and was considering treatment with pig cells, said,

The personality is situated in the brain. If one complements it with a very small number of cells from an aborted fetus, or from an animal such as a pig, then I don't think that it would disturb my overall conceptualization. But if we should talk about replacing half of the cerebrum, then we would be changing a large part of an individual's personality.

(Lundin 2002:337)

Worries about 'something alien taking control over their human self' were evident (Lundin and Widner 2000), and identity was located in different places within the body. At the same time, these interventions offered potential solutions to the 'identity crisis' associated with being chronically ill, the biographical disruption of illness (Williams 2003), and individual patients' pragmatic concerns to become healthy shaped 'personal morality', rather than wider ethical concerns. Moreover, in Lundin's analysis, patients stressed that 'their natural and real self can be restored with the help of technology' (Lundin 2002:342). In these accounts, the generative potential of the body, the possibility of it regenerating itself, is not considered; instead, they rely on notions of repair and replacement discussed already. However, reconceptualizing the body and identity as open ended and flexible reframes the ethical questions and implies fewer disruptive effects of cell-based therapies. Closure becomes unimportant, and, instead, 'embracing concorporeality' or 'multiplicity' becomes the primary aim. SC therapies as *productive* possibilities that mobilize the body's potential in a dynamic process of 'autogeneration', rather than reinstate a normative body, embody shifting and mobile subjects.

Towards a feminist bioethics of the body

The postmodern turn has thrown into question notions of the liberal-humanist subject, autonomy, rationality and the idea that identities can be read off from naturalistic bodies. Instead, identity has been seen as shifting, contingent and multiple. The body as a site of inscription and as a representational device has been reshaped, remoulded, and bodily boundaries have been increasingly understood as open, permeable, leaky. However, rather than implying that a postmodern view of bodies, self and identity undermines the significance of bodies, corporeality can be seen as central to subjectivity. We *are* our bodies, and the notion of 'biological citizenship' emphasizes this and points to a new kind of 'somatic ethics' (Rose 2007):

Ethics here are understood as a way of understanding, fashioning, and managing ourselves in the everyday conduct of our lives. If our ethics has become, in key respects, somatic, this is in part because it is our 'soma' – or corporeal existence – that is given salience and problematized – to some extent at least, our genome, our neurotransmitters – our 'biology'.

(Rose 2007:257)

Rose says our understanding of ourselves is increasingly biological and biomedical; we make sense of who we are, how to act with reference to bio-medical authority and other 'somatic experts' (genetic counsellors, bioethicists). However, this hardly implies a biological reductionism or a return to naturalistic views of the body; rather, biopolitics, as we saw in Chapter 1, implies a new kind of 'vital politics'. This goes beyond the 'identity politics' of earlier feminist projects.

> Feminist critiques of science, more specifically NRTs, have been underpinned by a concern that women have disappeared from view in the language and practices of research scientists, clinical practitioners and policymakers. It has been argued that research activities which have rendered women as experimental subjects, test-tubes or laboratories, and simultaneously given embryos a status and subject position outside women's bodies, have significantly undermined the status and subject position of women. For, in so far as the new technologies have encouraged the view that IVF and other techniques are dependent on technological expertise and know-how, women's bodies have been both devalued and written out of the picture. The notion that life, in embryonic form, may be sustained outside a woman's body transgresses previously accepted and understood boundaries. At the same time, the invasive approach of techniques such as egg collection, embryo lavage and insertion of drugs, instruments and other technical paraphernalia crosses boundaries in both directions by entering into a woman's body and taking parts away.
>
> (Kent 2000:170)

Feminist political projects have, since the 1970s, engaged with concerns about the biomedical exploitation of women's bodies and the gendered effects of technological developments. Women's bodies were marked as deficient, abnormal, chaotic and out of control. Differences were understood in hierarchical terms. Describing shifts in medical knowledge and anatomy from a view of the female body as deficient to different, Shildrick says,

> Where once the female was simply represented as a deficient version of the male ideal, the new knowledge constructed women as radically different from men, but no less made relative to and devalued against a male standard. In other words, difference was recognized but the hierarchy was maintained.
>
> (Shildrick 1997:31)

Furthermore, as indicated already, pregnancy, childbirth and reproductive technologies were sites for contestation and political action. Shildrick argues for a feminist bioethics that goes beyond consideration of aspects of women's health (or ethics of care) to develop a broader theoretical framing of ethical relations. Using heart transplantation as her example, she suggests that bioethics 'concerns itself with the unpredictable encounter or relation between one embodied subject and another where there is no clear distinction to be made between them' (Shildrick 2008b:33). By exploring the 'troubling' effects of heart transplantation and its implications for understandings of the self, embracing concorporeality and seeing the recipient as hosting the donated organ, she believes anxieties about an ontological threat to bodily integrity and the self might be allayed. However, within regenerative medicine, the distinction between one embodied subject and another seems to collapse further. If we see all bodies as 'chimeras' and assemblages, the mixing of cells and insertion of 'engineered' tissues or cell-based therapies can be understood simply as part of the leaks and flows that constitute bodies (embodied subjects) in the first place. Furthermore, as has been suggested, it becomes difficult to distinguish between reproduction and regeneration (Cooper 2008), and between reproduction of the person and of the 'soma'.

Conclusion

To conclude, therefore, I have reflected here on the extent to which contemporary life sciences destabilize modernist ideas about the subject. I have shown that the naturalized body has increasingly been called into question as bodies have become understood as *fabrications*, that is, they are made not born. Scientific, biomedical and cultural practices *materialize* bodies in multiple ways, and, far from being a 'skin-bounded entity', bodies are unstable, flexible and permeable. Bodily boundaries are hard to fix, and embodiment can be understood in terms of lived experiences of flux and change. Feminist theorizing on subjectivity has raised important questions for some time about liberal conceptions of a universal subject, and pregnancy has been seen as a paradigmatic case of 'split subjectivity'. The intermingling of bodies – the *chimeric* qualities of maternal and other bodies – unravels notions of bodily integrity or well-defined bodily boundaries (Martin 2010, Kelly forthcoming). Abortion and the collection and use of fetal tissue for use in research and therapies highlighted in diverse ways the ontological problems of relationships between the woman and the fetus. The form and meaning of fetal tissue were especially ambiguous and indeterminate, and fetuses are materialized in multiple ways in different contexts. Tissue-based technologies and tissue exchange imply new experiences of embodiment that are intercorporeal. Extracorporeality referred to an externalized and extended vision of the self and materiality of the body, whereas concorporeality referred to the co-residence of embodied subjects whose boundaries cannot easily be discerned. Nevertheless, in important respects, regenerative medicine could be seen as distinct from organ transplants or prosthetic devices, and, therefore, the insights from studies of how these impact on the identities or lived experiences of patient bodies were of limited utility in thinking through the meaning and significance of tissue regeneration. While we support calls for a feminist bioethics, how we think about the generativity of bodies and the effects of tissue- and cell-based therapies merit further consideration. They raise fundamental questions about life, death and what it means to be human, as we have seen already.

7 Life, death and immortality

Introduction

In this final chapter, I want to reflect on the ways in which we think about bodies, the meaning of embodiment and its significance for feminists, and the socio-ethical implications of emerging technologies that modify or transform bodies. In the first section, I will explore the issues around 'informed consent' as it relates to embryo collection. The second section briefly summarizes discussion around the ethics and politics of abortion and fetal-tissue use. The third section retraces the shift from transplantation medicine to regenerative medicine in order to generate new forms of biocapital in a bioeconomy characterized by social inequalities. Finally, drawing on feminist scholarship, I consider what a feminist bioethics can contribute to analysis of embodied futures.

I first had the chance to engage with the ethical issues around biobanks and explore why, and how, researchers valued biological samples as tools to investigate the linkages between genetics, lifestyle and health when researching the ethical protection afforded participants in the longitudinal epidemiological study 'ALSPAC' (also known as the 'Children of the 90s' study) (Goodenough *et al.* 2004, Williamson *et al.* 2004). Since then, my investigation of the science of TE and consideration of the policy debates surrounding the emergence of new 'tissue-engineered' products at national and international level (Chapters 3 and 4) pointed to deep-seated and long-running ethical controversies surrounding the procurement and use of human tissues for commercial gain and the institutional responses to these dilemmas. Such debates turned, crucially, on sustaining a separation of things and persons, protecting a moral order whereby persons (or the tissue obtained from them) are not treated as objects that can be bought or sold. International patent law and notions of 'intellectual property' relied on such distinctions, as do protests about the trafficking of persons, organs or body parts. Interestingly then, as I discussed (Chapter 3), one critique of the 'commodification' thesis is that those who seek to maintain such a separation unwittingly reproduce the very basis on which the capitalist economy depends (Hoeyer 2007). Rather, as we have seen, the transformation of capitalism has been linked to new, emergent life forms, new biotechnologies and the science of SCs and regenerative medicine. Biocapital represents a transformation of bodily potential to generate new forms of value, and the boundaries of bodies are seen as unravelling, becoming increasingly unstable and denaturalized. Bodies become *plastic*, 'fragmented', dismembered or disassembled. As tissues and cells are increasingly manipulated, reassembled and regenerated in the laboratory and *in vivo*, the notion of what it means to have, or to be, a body is reconstituted (Chapter 6).

Materialist understandings of bodies that naturalize or take as given their ontological status increasingly come under tension in analysis of regenerative medicine. Rather, recognizing that bodies are 'fabrications' (Shildrick 1997) made and born (Franklin and Roberts 2006) implies a reassessment of the connections between reproduction and regeneration. An unfolding drama has been taking place, as life itself and death are being 'remade' (Franklin and Lock 2003). The notion of 'living cadavers' pointed to the cultural production of 'death' (Lock 2002b, 2004). New scientific objects have been created, as the biological sciences have increasingly been directed towards realizing, or rather materializing, the productivity of bodies and multiplying life (Chapters 2 and 6). Projects to capitalize that surplus are at the heart of attempts to revitalize flagging neo-liberal economies and to invigorate developing economies in India, China and elsewhere (Rajan 2006, Salter 2006b, Salter *et al.* 2007, Cooper 2008, Bharadwaj and Glasner 2009, Patra and Sleeboom-Faulkner 2009). The life sciences are increasingly tied to the interests of business, as governments (including in the United Kingdom) seek to gain a competitive edge in markets for knowledge and to provide new solutions to contemporary problems.

The definition and prioritization of problems relate to changing population demographics and concerns about sustainable futures. The 'problems of old age' associated with longer life expectancy are heralded as justifying and creating a 'need' for cell- and tissue-based therapies; another reason is the shortage of organs for transplantation.[1] However, the political decision to prioritize regenerative medicine over other social and health needs is open to challenge. Investment in the speculative futures and promises of regenerative therapies remains patchy and uncertain, especially in the context of the global financial crisis. Such investment 'speculate[s] on the future of innovation' and risks a great deal (Kemp 2006a, 2006b, Martin *et al.* 2006, 2009, Rowley and Martin 2009). While the 'rhetoric of hope' mobilizes public support and patient expectations for what these therapies might deliver, clinicians and scientists themselves are embroiled in complex social interactions, the dynamics of which shape the possibilities for translation from research to clinical practice, which is not in itself a linear or unidirectional process (Chapter 2) (Martin *et al.* 2006, Wainwright *et al.* 2006b, 2007, Kraft, 2008, 2009, forthcoming).

In light of these developments, therefore, how far are we witnessing a transformation of social relations? To what extent are these emergent technologies truly transformative, or do they reify existing inequalities? I explored the gendering of the bioeconomy, the ways in which deeply entrenched gender subordination runs through, or mediates, the shaping of the scientific and industrial processes, which in turn have already produced, or are expected to produce, new therapies (Chapter 5). Women's position within the bioeconomy was examined, with women both as subordinated workers within the scientific community and as (potential) recipients of the benefits of regenerative medicine. Most significantly, however, the positioning of women relates to the connections between reproductive medicine and regenerative medicine. Waldby and Cooper's work (Waldby and Cooper 2008, 2010) on this is especially helpful in tracing the historical connections between women's exploitation under industrial-capitalist economies and drawing out ways in which contemporary patterns of exploitation are distinctive, suggesting that women's involvement, particularly within the SC industries, requires a different approach.

With reference to the work of other materialist feminist commentators, such as Donna Dickenson, they seek to redefine the meaning and importance of reproductive labour as *regenerative labour* within the SC industries and go beyond 'a Fordist industrial model of labour and the nation-state model of reproduction' (Waldby and Cooper 2010:10). In this reading, therefore, there is a need to distinguish between the forms of reproductive labour associated with women donating their body, say, for surrogate gestation of an embryo/baby, or donating ova for IVF treatment, and those women whose ova or embryos are used within the SC industries. They propose that, in the latter case, regenerative labour better describes the reordering of the female body's capacities:

> The organization of intellectually property in the life sciences recognizes the cognitive labour of the scientist and the clinician, but not the constitutive nature of the biological material or the collaboration of the donor [. . .] It is evident that the recognition of labour here is structured by a mind/body split, wherein the embodied productivity of the tissue donor does not figure. In [our] analysis, however, this embodied productivity is given a central place. Stem cell researchers require access to women's *in vivo* reproductive biology, the living interior processes of their bodies, as a generative site for biological materials. The donor's participation in this process is generally conceptualized as deliberative and contractual; that is, they participate insofar as they make an informed decision to consent to donation. However, we would contend that their participation should be understood as a thoroughgoing embodied collaboration that involves not merely the legal disposal of surplus biological material but rather the generative energies of the donor's biology, sustained over time.
>
> (Waldby and Cooper 2010:9; emphasis added)

Therefore, rather than suggesting that procurement of reproductive tissues for the SC industries is another form of reproductive labour, their argument is that:

> If different modes of technical production, scientific speculation and economic calculus call forth different capacities from the body, it is the very conception of what the body is capable of doing – the work it is capable of rendering and experimental systems in which it can play a part – that is under negotiation in the encounter between reproductive and regenerative medicine. While reproductive medicine demands a literal labour of reproduction from the female body, regenerative medicine is interested in the body's capacity for embryonic self-regeneration, prior to and independently of any process of development. Bodily potentiality is itself being reconfigured at the interface of new labour relations and the biological sciences.
>
> (Waldby and Cooper 2010:17)

Moreover, it is the temporal reordering of life that, in their analysis, distinguishes regenerative labour from reproductive labour.

Their argument is compelling and offers up important insights for consideration. Applied to the appropriation of aborted fetal tissue, then, this useful concept of regenerative labour, I suggest, needs further elaboration. As we saw earlier, two models of cellular therapy coexist: one draws on transplantation science, and the other on concepts of 'regeneration'. Where primary cells from fetal tissue are directly transplanted into the brains of those with degenerative disease (that is, as replacement cells), this practice is combined with expectations that these cells will *regenerate* the brain. Fetal tissue, once separated from the maternal body, has its generative potential exploited and reordered when it is deployed by the life sciences for cellular therapies, although the creation of fetal-cell lines has proved technically difficult.[2] The woman's productivity, embodied in the fetal tissue, is captured and reconfigured, as Waldby and Cooper's analysis of embryos and oocytes above suggests. Crucially, however, in Pfeffer's (2009b) analysis of the transformative effects of the labour of scientists on fetal tissue, she underplays the contribution of the women, that is, their labour. She over-emphasizes productive labour (the work of scientists and researchers) and neglects both consideration of any literal kind of reproductive labour and, more radically, whether valorizing women implies a more careful account of their exploitation and gender subordination. In so doing, she naturalizes women's bodies, by taking them for granted and operating within a framework of dualistic thinking that separates objects and persons.

What is needed, then, in extending and developing further a feminist analysis of regenerative medicine, is a feminist bioethics that takes account of these innovations. Therefore, as I began to explore in the last chapter, this implies a reworking of concepts of self and subjectivity in the context of women's (changing) relationships to their bodies. Moreover, it implies a deeper consideration of the meaning of key ethical terms that currently frame and underpin women's (and men's) relationships to these technologies, and a reassessment of the significance of *embodiment* for feminist theory. In this final chapter, I attempt to draw together the threads of the book and examine the claims made for regenerative medicine to prolong life and address the problems of ageing bodies. A critical assessment of such claims and the normative assumptions underpinning the innovation and regulation of new tissue- and cell-based therapies is presented. I reflect on how regenerative medicine 'remakes' life and death (Franklin and Lock 2003). The links between reproductive medicine and regenerative medicine are examined in relation to 'embryonic economies' (Franklin 2006a), and this is contrasted with the fetal-tissue economy (Kent 2008a). Whereas embryos commonly symbolize 'life' and have been at the centre of disputes about personhood and the patenting of life, I suggest that fetal tissue symbolizes both life and death. Fetal 'death', in the case of abortion (or miscarriage), releases the potential for life when tissue is donated for research or therapies. Whereas transplantation medicine redefined how the 'death' of a person was understood, regenerative medicine reframes ideas about the vitality of living tissues and cells. Commercial exploitation of tissue- and cell-based therapies frequently emphasizes their capacity to overcome the degenerative effects of ageing and even offers the promise of 'immortality', or at least of extending life (Hall 2003). In conclusion, the chapter considers whether the political claims made for regenerative medicine are likely to have transformative effects and asks what kinds of future are embodied in tissue- and cell-based therapies, and to what extent are the promises and expectations surrounding them likely to be realized.

Women and embryos: informed consent

Embryo collection is not a new practice, and its cultural significance cannot be underestimated. Describing early practices of embryo collection, anthropologist Lynn Morgan notes that embryologists were mostly scientists:

> who lacked access to the women whose bodies harboured the coveted specimens. They had to depend on the kindness of their clinical colleagues for access to specimens. Embryo collecting was born, then, as a collaborative effort between research scientists, clinicians, and pregnant (or formerly pregnant) women.
>
> (Morgan 2009:6)

It was facilitated by the medicalization of pregnancy loss (ibid.:51) and began at a time before embryos or fetuses acquired attributes of 'persons'. The extent to which this might be considered a 'collaborative effort' is an issue to which I will return; most significantly, scientists created an 'embryological origin story' and an entity called 'embryo' (ibid.:7), or what Hopwood calls an 'embryological view of life', which in turn, Morgan suggests, contributed to the naturalization of 'the embryo' and normalization of collection practices. Embryos, then, are cultural products that emerged from these practices. Anthropological and historical study of embryo collecting provides important insights into how embryological knowledge was produced. The Carnegie 'stages' of human development were constructed from the meticulous study of *dead* embryos and fetuses obtained through various means. At the beginning of the twentieth century, first-generation embryo collectors such as Franklin Mall, anatomist and embryologist at Johns Hopkins University, relied on personal and professional contacts for the procurement of tissue. Then, in processing these 'dead specimens', he, and others at that time, such as the German embryologist Wilhelm His, thinly sliced them and placed them on slides, materializing the embryos, giving them body and form and creating wax, plaster and, later, plastic models (Hopwood 2000, 2005, Morgan 2006, 2009). The founders of the Carnegie collection collected hundreds of late-term fetuses, and the reputation of the collection has frequently relied on a separation of the messy corporeal work of dissecting dead fetuses, to create these sections and models, and their more recent transformation into highly valued and visualized 'icons of life', which can generate research grants and funding (Morgan 2004, 2009).

In scientific terms, the value of the Carnegie collection is difficult to measure. It is part of a larger collection of 10,000 embryos held by the National Museum of Health and Medicine, Washington, in the Human Development Anatomy Center. It is a point of reference for medical students and SC scientists today. The collection, which divided human development into twenty-three stages, is itself a cultural achievement, as Lynn Morgan has eloquently argued. What emerged was a distinction between the 'embryonic period', defined as the period lasting up to completion of organogenesis, the formation of the organs of the body at around 6–8 weeks' development, and the subsequent fetal stage. In other contexts, less well-known collections such as the one in the medical museum of Copenhagen, which I visited in 2007, have received relatively little public attention, but, similarly, the tissues were obtained from the hospitals where women came for assistance and clinical attention.[3] Commonly, details of the circumstances of the women were not recorded, and women were written out of the picture in 'embryological narratives' (Morgan 2009). Embryos have their own 'social lives' and acquired an ontological status independent from women (Appadurai 1986, Franklin and Roberts 2003).

In more recent years, changes in the biomedical-research infrastructure and shifts towards capitalist systems of exchange co-produced embryos and embryonic material as at the boundaries of the distinction between commodities and personhood, for

> Embryonically derived material and other types of reproductive tissue slowly acquired traits of personhood in conjunction with changes in the

institutional context and the exchange systems it became part of. In particular this became apparent as the exchanges took the shape of the commercial exchanges related to the capitalist ideology (where the object is given a price).

(Hoeyer *et al.* 2009:15)

In other words, a new moral order emerged, and entitlements to embryonic (and fetal) material were reconfigured.

One thing that is striking about Morgan's discussion of embryos as 'icons of life' is the way in which, in the North American context, the terms embryo and fetus are used interchangeably. Paradoxically, embryological knowledge about 'life' depended on the collection of 'dead' embryos.[4] In contrast, in the British context, distinctions between embryos and fetuses have become institutionalized within the legal and regulatory framework and draw on a specific scientific discourse that emphasizes both the stage of development post-fertilization and the geographical location of the entity (Chapter 4). This was evident at a seminar where Nick Hopwood presented an account of the work of Wilhelm His and showed pictures of the 'embryos' he worked with, when a clinician in the audience, a fertility expert, responded by insisting that it wasn't an embryo but a fetus, because of its age and because of its biography (Hopwood 2000, 2007). In Britain, 'embryo' has come to signify the pre-implantation, *in vitro* embryo or blastocyst created in the fertility clinic, the scientific object of embryo and much (though not all) SC research. As we saw earlier, since 1990, it has commanded separate regulation and an entire regulatory agency, the HFEA, devoted to its protection and management. The different political and cultural context in Britain, which enabled embryos to be separated from abortion politics, secured the scientific enterprise in ways that were not achieved in North America. This highlights the need for investigation of the specific features of 'embryonic economies' and 'fetal tissue economies' in different sociopolitical contexts, and recognition of the ways in which diverse tissue economies draw on, and mobilize, distinctive and overlapping meanings of human biological materials.

Sarah Franklin's important contribution to our understanding of the interface between reproductive medicine and regenerative medicine has illuminated key features and provided us with insights into the genealogy and social life of embryos (Franklin and Roberts 2003, Franklin 2006a, 2006b, 2007a, 2007b). She and others contributed to the development of standardized consent procedures for embryo collection in British IVF clinics (Franklin *et al.* 2008). In Britain, the embryo supply is a strictly regulated, managed economy, and women who donate embryos are frequently seen as committed to advancement in infertility treatment and therefore likely to want to donate embryos for research (Kent 2008a). There are, however, complex views about which embryos might be regarded as 'spare' or available for donation to research, and not all those undergoing treatment are willing to donate (Parry 2006). Moreover, it has been suggested that there is frequently elision between 'egg sharing' for research and that for therapeutic use (Roberts and Throsby 2008). Some scientists considered it more ethically acceptable to use embryos obtained following prenatal genetic diagnosis (PGD) than 'spare' embryos from IVF treatment. PGD embryos were 'genetically affected' and, therefore, were deemed not suitable for implantation in a woman and so more available for research. Such 'spare' embryos could be seen as being 'rescued', as they would otherwise be 'left to perish' or 'wasted' (Wainwright *et al.* 2006a).[5] It has also been debated whether consent models should enable donors to place restrictions on the use of embryos and lines derived from them, for therapeutic use, or for gamete derivation (Ehrich 2010).

Despite emphasis within the regulatory framework on individual autonomy as a basis for informed consent, evidence suggests that, in PGD, where women/couples are asked to choose which embryos to use, an ethical model based on relational autonomy is needed, as staff shape and frame the kinds of information given in order for reproductive choices to be made (Ehrich *et al.* 2007). Ethical models that are individualistic fail to engage with both relational aspects and the global ethical dimensions.

In Britain and in Europe (as discussed in Chapter 3), regulation of embryo and other human-tissue research has increasingly centred on the guiding principle of informed consent (Kent and ter Meulen 2011). This has been problematized, because it relies on neo-liberal assumptions about autonomy and consent as key to ethical regulation. For example, Widdows elaborates how *exploitation* is a central concept, and that, although, for example, women might consent to donate eggs (and, I suggest, fetal tissue) for use in research or therapies, this isn't necessarily morally acceptable or socially desirable – that is, they do so in conditions of gender subordination, where the circumstances in which they make choices and give consent are exploitative (Widdows 2009, 2010) (see also Wilkinson 2005, Dickenson 2007). Drawing parallels between prostitution and egg 'donation' (or selling), Widdows argues that:

> both trafficking for prostitution and egg donation for research are exploitative and thus should not be endorsed by feminists. Moreover, the failure to name such practices as exploitative serves to normalize and extend them, thus leading to the exploitation of more women.
>
> (Widdows 2009:6)

Following this analysis, to assume that consent, as a means of transferring bodily material to researchers, is given freely misses the significance of the conditions under which such 'choices' are made. Those are conditions of gender subordination, and consent does not, of itself, cancel out exploitation. Hence, the international debacle surrounding the Korean researcher Hwang's fraudulent claims to have successfully 'cloned' human SC lines[6] illustrates the point that women in his laboratory and elsewhere were exploited as part of a nationalist project (Gottweis and Kim 2010), because (a) their health was put at risk and (b) because asymmetries of power characterized the relationship between the women (some of whom were junior researchers) and senior scientists. It also drew attention to the lack of international consensus around ethical standards, and, since then, the work of organizations such as the International Society for Stem Cell Research (ISSCR)[7] has been directed to promoting these (Salter 2005, 2007a, Salter and Salter 2007, Waldby and Salter 2008). Moreover, as I noted previously, opponents of egg donation across the world have vociferously argued that markets for eggs (for therapies or research) exploit poorer women, for whom financial inducements represent a chance to relieve financial hardship (Beeson and Lippman 2006). Tissue exchange, then, must be understood as a situated and contingent practice embedded within a gendered global political economy. This is a bioeconomy, where women's subordinate position results in disparities between the value accorded their reproductive and regenerative labour and, following from this, an affront to dignity. However, rather than arguing for a 'fair price' or recompense (for prostitution or eggs), or being fearful of neo-colonialism, Widdows (2009) suggests we need to return to fundamental concerns about gender inequalities and acknowledge that choices women are given are not always evidence of empowerment or greater freedom. Being asked to choose between different kinds of 'bad choice', for example, poverty or prostitution/egg selling, is not of wider benefit to women, even where they end up wanting these choices (Widdows 2010). This means questioning the nature of the choices women make when donating eggs, embryos or aborted fetal tissue for research, and the limitations of a contract based on consent. Women can be viewed as *co-labourers*, but voluntaristic accounts of 'collaboration' are undermined. In so far as collaboration recognizes women as co-labourers, it is a useful concept, but it does not adequately highlight the exploitative nature of such a collaboration where it is regarded as a partnership, or women's contribution is assumed to be as equal with scientists who procure and use reproductive tissues as research tools. So Waldby and Cooper's use of the term 'embodied collaboration' draws attention to the different contribution of women's bodies and body parts, which constitutes regenerative labour. However, Morgan's description of the 'collaborative effort', which features in early embryo collecting, is less illuminating.

Abortion, fetal death, corpses and organ donation: the right to choose

Feminists have also debated at length the rights of women to abortion, and pro-choice arguments have relied on liberal concepts of autonomy and freedom. Although I support the right to abortion, rights-based claims are, of course, theorized differently by various commentators (e.g. Lee 2003, Grear 2010). What were important in the study investigating fetal-tissue use in research, which I described above, were the ways in which regulatory and scientific practices framed women and their fetuses (Pfeffer and Kent 2007). As we saw, in Britain, abortion law does not currently acknowledge women's rights; rather, it frames abortion as a public-health issue and as needing supervision and management by the medical profession (Sheldon 1997).[8] Abortion is medicalized, and access to it requires women to subjugate themselves to the power of medical practitioners and to make claims that damaging social and health consequences will result if an unwanted pregnancy is not terminated. The earlier guidelines on fetal-tissue use in research were characterized as paternalistic and as implying that women needed protection from themselves as well as unscrupulous researchers. The separation principle – separating women's decision about having an abortion from the decision to donate the tissue, and separating the researcher from the woman – underpinned the guidance and shaped practices. More recently, following the UK 2004 Human Tissue Act and, five years later, a new code of practice for researchers (Human Tissue Authority 2009), the emphasis on informed consent superseded the earlier approach. In the interim period, when this research was being conducted, we found evidence of confusion and variation in the extent to which women were informed about the research uses of the material and uncertainty about whether consent should be specific or generic. Of particular interest were the distinctions made between research use of fetal tissue and that of pre-implantation embryos, even where they could be found in the same laboratories, being used by the same scientists.

Crucially, for scientists, they were both human biological materials and as such were (at least for some) in certain respects of similar ethical significance. However, the potential embodied in pre-implantation embryos to be implanted in a woman and to develop into a baby meant they were afforded greater ethical protection, and their totipotent characteristics meant they were especially highly valued as research tools. Embodying the 'embryological world view' as symbols of life, they were prized 'icons of life', and their promissory value was high. At the same time, however, scientists asking different research questions, who regarded a developmental approach to understanding SCs as important, valued the special qualities of fetal tissue, which could yield primary cells at early stages of differentiation. Some scientists remarked on what they perceived as a paradoxical contrast between the 'light-touch' regulation of fetal tissue obtained following a highly contested and contentious practice of abortion and the heavy-handed, bureaucratic protection of embryos created as part of IVF treatment (Kent 2008a).

Cultural practices and attitudes towards death are framed by diverse ethical values. In relation to abortion politics, of course, a central debate is the claim to personhood of the fetus and, following from this, whether or not it has rights separate from, or independent of, the woman's maternal body. In the British context, we saw that there are contradictory ideas about fetal tissue, and the status of fetuses is highly ambiguous. I argued that fetal tissue was viewed as both waste and as a corpse (Kent 2008a).[9] On the one hand, it was seen as like other tissues or body parts of the woman, which meant she could give permission for a termination to take place (like any other surgical procedure), and it could be disposed of as clinical waste, or donated for research. However, it was also regarded as special, and different from other types of human tissue, and, once aborted, it could be formally recognized as a corpse, deserving of respect and burial or cremation. The shifting meanings and understandings of 'the fetus' as waste, corpse, research tool, baby, body part of the woman revealed the different ways in which fetuses were materialized and their unstable ontological status (Williams 2005). The possibilities associated with both transplantation and the revitalization or reanimation of tissue using cell-culturing techniques meant that, following abortion, 'a fetus' could be seen as symbolizing both life and death *simultaneously*. As liminal objects at the margins of life and death, the appropriation, redirection and reuse of fetal tissue for research or clinical applications realize surplus value in specific ways that are both similar to, and different from, those relating to pre-implantation embryos.

Similarly, Lock's 'living cadavers' – the ventilated but not quite dead bodies (heart-beating corpses), from which organs can be 'harvested' for transplantation – embody both life and death (Lock 2002b, 2004). However, in important ways, transplantation medicine and regenerative medicine differ, as we saw previously: organ transplants are revived and remodelled to provide *replacement* parts, whereas *regenerative* medicine expands and amplifies tissues, engineering and generating new tissues (Cooper 2008). This is the basis of the claim that regenerative medicine represents a 'new paradigm', a fundamental shift in approach to clinical intervention and treatment.

From transplantation medicine to regenerative medicine: innovation stories

A number of important and interwoven strands of historical significance can be identified in constructing accounts of innovations in contemporary life sciences and clinical medicine. In cell biology, the development of cell-culturing techniques (Chapter 2) and other methodologies were critical in furthering understanding of cancer cells and concepts of 'immortality' and provided the tools for extracting and identifying cell types and culturing cell lines. The 'science of monstrosities' (teratomas) was important historically for SC science, and, preceding this, long-running debates that sought to theorize and explain the process of epigenesis also underpin more recent theories of tissue 'regeneration' (Cooper 2003, 2005, Landecker 2007). The post-war collection and transfusion of blood also contribute to contemporary accounts of tissue exchange (Brown and Kraft 2006, Brown *et al.* 2006b). Scientific study of leukaemia and clinical developments leading to treatments such as bone-marrow transplantations have also been important for SC biology and have recently been redescribed as the first success story of SCs. However, additionally, they have also provided a cautionary tale about the 'dark side' of SCs, captured in the theory of the cancer SC having 'an ability to wreak well-regulated havoc, undetected by the body's immunological surveillance mechanisms' (Kraft forthcoming). Moreover, as Alison Kraft notes, bone-marrow transplants for leukaemias are seldom curative, and, rather than an uncritical acceptance of the potentialities of SC therapies, she shows that there is a 'double edge' to them, a potential for danger, so that, 'in an era seemingly entranced by the idea of cellular therapies, the shadow cast by the cancer stem cell serves as a signal for caution in the future clinical use of stem cells' (ibid). Furthermore, despite the significant commercial investment in them, the future commercial potential of SC therapies is also uncertain (Martin *et al.* 2006).

Since the days of early embryology, new reproductive technologies gave us the *in vitro* pre-implantation embryo and established clinical institutions and practices facilitating the production of embryos and egg procurement through access to a population of young women. Fetal medicine created the fetus as a new type of clinical (and scientific) work object and a new patient (Casper 1994, 1998). While the 'new genetics' knowledge led to a focus on bioinformatics and insights into molecular life, exchange of human biological materials between bodies has increasingly become mediated by diverse and complex processes of technical manipulation, reconfiguration and reassembly, embedded within a sociopolitical, institutional and regulatory context that is also complex.

Umbilical-cord-blood banking has also been controversial for its claims, particularly by private banking services, to offer potential, as yet unidentifiable, therapeutic benefits, trading on the desires of new parents (especially mothers) to protect the future of their child's health by banking SCs obtained from the blood sample, in case they are needed at a future date. The speculative nature of this service and its redirection of potential donors to public banks, which use donated cells for allogeneic transplantation, have been criticized on a number of counts by health professionals and others (Busby and Martin 2006b, Busby 2010a). We are also witnessing the proliferation of other 'banking' services – banking of menstrual blood and placental tissue.[10] Although the therapeutic benefits of public cord-blood banks are generally widely accepted, the potential benefits to a child of its own cord blood are very uncertain, indeed unlikely, given the current state of knowledge.[11] Moreover, even though evidence of the therapeutic benefits of SCs from menstrual or placental tissue is very weak, such services trade on the promissory value of hopes and expectations, of speculative futures.

As we saw in relation to the techniques using somatic cells such as ACI or engineered skin (Chapters 1, 5 and 6), both technologies that source cells from a single patient for his/her own use and those that profilerate allogeneic cells to create an 'off-the-shelf' engineered product (or cell therapy) potentially reconfigure bodies, self and identity. These technologies are shaped by political and economic interests that impact on clinical practices and the distribution and accessibility of treatments. Moreover, although the focus of ethical debate and controversy has often been in relation to hESCs or the use of gametes in research (for example, in SCNT) and therapies, *all* human-tissue use in Britain and in Europe has recently been at the centre of ethical controversy, not least because of its potential exploitation for profit. Whereas the drive for innovation has been tied to concerns about the unmet demand for organs and intractable degenerative diseases, public-health discourse frequently appears in tension with the drive for new ways of generating wealth. This, in turn, has been evidenced by the relatively unregulated proliferation of companies trading on the hopes and expectations of unsuspecting consumers of unproven therapies, as well as formal disputes around the patentability of new 'inventions' (see below). While organs became 'therapeutic tools' worked on by clinical experts to shape and prepare them for reanimation in a recipient body, the degree of 'manipulation' or 'engineering' employed to transform other tissue and cell products renders them subject to new types of property claim (Hogle 1996, 1999, Dickenson 2007, Hoeyer 2007).

Ways of generating and creating biovalue now extend beyond the transplanting of whole organs, whole blood from one body to another, or the bodily transactions involved in exchanging gametes characteristic of reproductive politics (Waldby 2002a, 2002b, Kent *et al.* 2006a). 'Gift economies' have been a focus of critical reassessment, as gift relationships offer only limited characterization of the contemporary global bioeconomy (Waldby 2006, Waldby and Mitchell 2006, Busby 2010b). Rather, intensification of the process of capitalization and transformation of capitalism to generate newer forms of biocapital reposition persons, bodies, body parts, tissues and cells. Nevertheless, there are continuities with the past; gender inequalities are not, it seems, swept away in the new regenerative medicine. Moreover, evidence suggests that other social inequalities, including racism, will be perpetuated, although I don't examine these in any detail here (Skloot 2010). The debate around egg selling, like that around organ trafficking, highlights that the direction of travel of human tissues is more usually from the poorer developing economies to wealthier citizens in the 'First World', and, in the United States, black populations benefit less from organ-transplant services or the market for body parts (Goodwin 2006).[12] At the policy level, efforts to protect the vulnerable while promoting organ exchange and biotechnology-industry initiatives run the risk of getting embroiled in ethical deliberations that extend beyond the regulatory remit of international agencies (Chapter 3).

Cultural and ethical diversity has been valued within Europe, both constraining and shaping debate around the regulatory implications of regenerative medicine. While the principle of subsidiarity has been applied to enable ethical decisions to remain a national concern, inconsistencies within a market-driven approach have occurred. For example, there are continuing uncertainties about the extent to which different member states might be willing to challenge the EMEA approval of cell-based therapies (if, and when, they come on stream) by not permitting those derived from nationally restricted sources (e.g. hESCs) to enter national markets.[13] At the same time, patients' expectations about such therapies create demand for therapies that are not yet shown to be clinically safe or effective. Indications are that there is already widespread selling of SC therapies (Lau *et al.* 2008, Glasner 2009, Patra and

Sleeboom-Faulkner 2009). A recent EMEA press release emphasized that patients should be aware that many such treatments for multiple sclerosis, spinal injuries, stroke, heart disease, eye disease and other degenerative diseases are not approved for clinical use, except as part of clinical trials or special exemptions for compassionate use (European Medicines Agency 2010). At the time of writing, since the new RATMP came into force in 2007 (see Chapter 3), no *SC* therapies have yet been approved for use in Europe under these regulations. (Chondrocelect, an autologous *somatic* cell treatment for repair of knee cartilage, was one of the first ATMPs given marketing authorization.)[14] Nevertheless, press coverage provides evidence of patients travelling abroad for SC treatment in recent years, and the UK Multiple Sclerosis Society and UK Parkinson's Disease Society have cautioned patients about seeking unproven SC treatment outside authorized clinical trials.[15] As we saw earlier (Chapter 1), the first alert ever issued by the UK HTA in 2006 concerned the distribution by companies of SCs that were intended for research and not human application (Human Tissue Authority 2006).[16] The ISSCR also provides advice for patients, noting that claims made for SC treatments are widespread and often not supported by scientific evidence, nor properly controlled. It is currently compiling a list of clinics and service providers claiming to offer SC treatments and intends to ask such clinics to supply evidence of ethical approval for the treatment and supervision by a regulatory authority.[17] An ISSCR Taskforce noted that, 'the problem of misleading direct-to-consumer advertisement of stem cell therapies has become all too familiar' (Taylor *et al.* 2010:1). Their report highlights the significance of direct-to-consumer advertising of such therapies and, by implication, the limitations of current regulation to police such activities. As a medico-scientific society, ISSCR recognizes that the approval or licensing of such activities is beyond its remit or responsibility, but its actions are motivated by the desire to protect the reputation of good science and the potential for appropriate translation of SC science into the clinic in the future.

In short, the development of cell- and tissue-based therapies is seen to have different implications for diverse social groups. Those who are vulnerable, marginal and sick are more likely to be co-opted as egg sellers or (mis)sold unproven treatments. There are disproportionate costs to women for the procurement of reproductive tissues. There is potential to exploit the disadvantaged through diverse forms of 'bioprospecting' within the newly configured 'bionetworks' or assemblages of technology and politics (Ong and Collier 2005, Patra and Sleeboom-Faulkner 2009). State-sponsored development to gain competitive advantage is evident, for example:

> A bioethical regime, driven more by the imperative to develop a life-science economy than by the ethical concerns of varied ethnic groups, woos foreign biologists who believe that, in Singapore, there are fewer constraints on research and the best opportunities for career development anywhere.
>
> (Ong 2005:343)

A global moral economy maps on to investment in the new life sciences (Salter 2007a, Salter and Salter 2007, Waldby and Salter 2008, Sleeboom-Faulkner and Patra 2009), and new forms of ' biological citizenship' are in evidence that emphasize both individual and collective valuations of biological life and active participation in technological futures (Rose and Novas 2004, Rose 2007). The role of bioethics as a kind of technical ordering that assists in governing bodies has been criticized for legitimizing scientific and clinical practices and for offering 'a clean bill of health' to practices that deserve more critical review (Rose 2007).

Feminist (embodied) futures?

In an assessment of the UK HFE Act 2008, Marie Fox suggests that feminists were not vociferous enough in expressing concerns about the revisions to the 1990 Act, and an opportunity to discuss what it means to be human was missed (Fox 2009). As I have already argued (in Chapter 4), powerful political interests shaped debate and determined the direction of change and support for the scientific enterprise by the British government. Of particular relevance to my discussion were the bracketing out of proposed changes to the timing of abortion and permissions given to create 'human admixed embryos', which were, on occasion, represented as in women's interest because they reduced the potential demand for human ova – although this is far from clear cut. (Professor Alison Murdoch at Newcastle University and others argued there was a research need for human ova.[18])

At the centre of debate about the ethical significance of 'admixed human embryos' was concern about the implications of interspecies mixing for what it means to be human. Animal–human boundaries could be seen as threatened by the proposed mixing of animal eggs with human somatic cells. However, the passing of the Act to permit such mixing *for research purposes* (distinguished from reproductive or therapeutic use) judged that human subjectivity was not threatened by such practices, so long as interspecies embryos were never implanted in a woman's body. The moral divide, which sought to preserve or sustain 'humanness', here operated in terms of embryo implantation *within a woman's body*; violation of what it means to be human could, it seems, only occur if such an 'admixed human embryo' was gestated by a woman. The notion that such entities were in any sense 'human' was the outcome of political deliberations that, I argued, sought to bring such practices under the control of the HFEA as a regulatory agency. The very idea of chimeric entities being created in this way poses very interesting questions for feminists, such as: does this reinforce the significance of (female) embodiment and gendered bodies? Does it reaffirm the value of reproductive labour? How are women implicated in this reasoning relating to chimeric embryos? I find Susan Kelly's work on fetal microchimerism illuminating and helpful to think through some of these questions (Kelly forthcoming).

Feminist readings of the significance of the maternal body have been important for emphasizing the features of connectivity and for elaborating notions of relational autonomy and an ethics of care. Kelly's study, which explores how fetal and maternal cells cross the placental 'barrier', mixing with those of 'the other', presents a view of *all bodies* as chimeric. This mixing has been a focus of investigations within fetal medicine and seems radically to reconfigure both the fetal and maternal subjects. Indeed, in an astonishing interview with a pre-eminent fetal-medicine clinician, he told me and my co-investigator of this phenomenon and his desire to cut open his own mother's body to establish which of her four sons' cells were dominant! Whether for dramatic effect, designed to shock us, or not, this image conjured up a bizarre vision of the ways in which techniques for identifying cells of 'another', a genetically distinct individual, might be located within a maternal body. Moreover, as Kelly explains, the mother's cells cross the placenta in the other direction and inhabit the body of her offspring. Such mixtures unravel, in previously uncontemplated ways, the meaning of bodies, their significance for identity and subjectivity (Kelly forthcoming, Martin 2010). This same clinician was collecting fetal cells *in utero* at the time of pregnancy termination for research, which in itself seemed to raise challenging questions about the direct access researchers had to maternal bodies to collect research materials in this way (Kent 2008a). So, while embodiment has been an important concept for materialist feminists and continues to have significance in revisionist accounts, such as that by Waldy and Cooper, of 'regenerative labour' as exploiting 'embodied productivity', its meaning deserves further elaboration.

Feminist (techno)science studies afford extensive insights into the unravel-ling of bodies as they become disaggregated 'bits of life', and the boundaries between 'nature and culture', human and nonhuman become blurred (Smelik and Lykke 2008).

> The new biotechnological discourses bring to the fore the material foundations of the embodied self, including its biological and genetic material. The emphasis on life marks a shift away from the deconstruction of layers of textuality, and toward an understanding of the inextricable entanglement of material, biocultural, and symbolic forces in the making and unmaking of the subject.
>
> (ibid.:xiii)

However, rather than bodies being understood as bounded entities, 'bits of life' refers to 'infotechnological and biotechnological processes of bodily redesign and convergences [. . .] the blurred boundaries between organism and technology and between matter and discourse' (ibid.:13). Following Haraway, bodies are reconfigured and transformed by contemporary 'technobiopower' (Roberts 2008). Roberts's study of hormones as the 'messengers of sex' shows how sexed bodies are being reconfigured. Rather than being viewed as a homeostatic system, controlled and regulated by hormones, she argues that bodies are situated within 'fluid ecologies', characterized by flows of hormones from, and between, material bodies and the environment (Roberts 2008). Such insights have salience for thinking about the flows and exchanges of cells and tissues from bodies, and the disassembly and reassembly of bodies in contemporary bioengineering and regenerative medicine.

Feminist legal scholarship is also helpful in illuminating the issues at stake in thinking through the meaning of embodiment and its significance for personhood, rights-based claims and legal subjectivity. For example, high-profile legal cases such as the one brought by Diane Blood against the HFEA shed light on the way in which legal discourse produces feminist subjectivity/ sexed subjects. The 1990 HFE Act has been criticized for its normative heterosexuality, which, to an extent, was revised in the recent, 2008, reform of the Act (Keywood 2000, Fox 2009). Blood's request to doctors to obtain and store sperm from her dying husband, followed by a rights-based claim to the HFEA to use the sperm, highlighted her positioning within a heterosexual economy. Her supporters, Keywood argues, naturalized her desire to have a child of her husband's, but, contradictorily, her request also revealed the fictitious nature of such a claim in so far as she was not infertile, her husband was dead, and the child would be legally fatherless.

> Through Blood, the discursive formation of the subject is exposed and the very instability of the category 'Woman' is made explicit. The potential for transgression from unviable to established gender identities reveals itself in a legal system which is incapable of fixing the subjectivities it purports to define and create. The threat posed to the culturally intelligible domain, occupied by those deemed worthy of reproductive technologies under the 1990 Act, by the abjected gender identities deemed unworthy of parenthood, forces the legal system to acknowledge its inability to contain and sustain the boundaries of viable corporeality.
>
> (Keywood 2000:331–2)

We see here that legal attempts to fix subjectivity come undone and can be deconstructed. Postmodern understandings of the body, identity and subjectivity provide an important basis for uncovering these representational acts. At the same time, we see that no body is pre-discursive; rather, that legal subjects are *productions* that constitute bodies and embodiment in specific ways. So, in a critical review of liberal legalism, which relies on a universalist and disembodied subject, uncertainties about the legal significance of bodies and embodiment come into view.

> Biotechnical innovations, such as gene splicing, cloning, and biologically-derived medicines which fragment the body into units of information – intellectual property – point to a general postmodern instability of the body's image. In this shifting legal topography, the corporeal location of

the rights-bearing subject is uncertain, for it is no longer clear exactly what or where the body *is* in scientific, legal and cultural terms.

(Halewood 1996:1349)

A number of conceptual and practical difficulties emerge when we see cell and tissue therapies as destabilizing the meanings of bodies. Bodies seem to disappear as they are denaturalized, and the significance of gendered bodies also begins to unravel. Conceptualized as chimeras, as a cellular mixture, whether following pregnancy or following tissue/cell transplants, bodily boundaries (and notions of bodily integrity) collapse. There is, then, no place for a foundational basis of epistemiology or ethics that is premised on essentialized ideas of bodies. Ethics, as the intersubjective agreement of how social life is ordered, poses questions about the moral economy of neo-liberal capitalism for feminists that goes beyond, but I suggest also recuperates, gendered bodies. This is a task that extends beyond the scope of this book and is itself a distinct, though related, project. My question is, to what extent does feminist argument depend, not only on gendering the body and re-embodiment, but also on prioritizing the ethical status of reproductive tissues – embryos, oocytes, amnion, placenta, fetuses and cord blood?

The dilemma for feminists, as I see it, is to sustain a reading of the gendered body as important for situating women within the bioeconomy. The ways in which reproductive tissues are mobilized and made available for use by researchers and corporations are both a material and a symbolic disruption of bodily integrity, in so far as they lay open and recapture, or exploit, the leakiness and fluidity of bodies. Yet, if we are to avoid reliance on a critical argument, which sees human tissues as increasingly commodified, because it turns on an object–subject dualism, we need to elaborate a new moral economy that engages with the meaning and significance of embodiment. It has been suggested that health-care law has conceptualized the body in three ways: 'as an object of choice, a site of property and a source of vitality', which may be useful for conceptualizing legal embodiment (Fletcher *et al.* 2008). As we have seen already, consent for donation of human biological materials is a key plank of the legal framework for tissue regulation and highlights the choices individuals make in this regard, but the institutionalization of choice and the rhetoric relating to its practice deserve, and have attracted, close scrutiny (Fletcher *et al.* 2008). In my earlier discussion of fetal-tissue donation, I noted that the choices women might be offered were delimited by the contradictory and ambiguous meanings attached to the woman's relationship to her fetus and enshrined within the Polkinghorne code of practice and subsequent HTA code (Pfeffer and Kent 2007). This act of donation is also construed as giving up any property rights in the tissue and making available its generative potential. The challenge, then, is to recuperate legal embodiment in order to expose the injustices of corporate claims to property rights and to profiteering from exploitation, while recognizing that 'the pre-market body is a fiction' and that liberal legalism is inadequate to account for the discursive production of bodies and embodiment (Halewood 1996). This isn't necessarily to suggest that new and emerging cell-based therapies have no public benefit. Rather, this implies a public-health ethics, expressed in health-care law, where the contributions of diverse social groups are equally valued, and the benefits of these technologies are widely shared (Dickenson 2007, Hayden 2007).

Conclusion

This book has sought to track a number of important scientific and clinical developments that can be broadly described as 'regenerative medicine'. In the process, I set out to synthesize and present findings from research carried out to investigate the governance issues surrounding the emergence of new tissue-engineered products and to explore the use of aborted fetal tissue in SC science. The central themes in my analysis of this research have focused on the ethical sensitivities raised by these technologies and the implications for understandings of what it might mean to 'regenerate bodies' and for gender relations. I have shown that evidence of attempts to commercialize tissue and cell therapies is patchy, and, despite strong government support for research in this area, there is much more to be done before these therapies can be expected to reach the clinic or become widely available. Critical appraisal of the claims that regenerative medicine represents a 'new paradigm', I argue, needs to address questions about both whether it will deliver health benefits and also whether or not it has transformative potential. It seems that these technical and practical possibilities do imply new kinds of relationship between our selves and our bodies. The symbolic significance for meanings of embodiment, for identity and subjectivity merits close attention, as bodily boundaries are increasingly shown to be leaky and unstable. However, at the same time, rather than concede that postmodern subjectivity undermines the importance of gender inequalities, I am concerned that we recognize how gender inequalities are perpetuated through tissue- and cell-based technologies.

Women's participation in the scientific enterprise, I argue, is constrained in specific ways: they are most frequently positioned as junior researchers and, in the clinic, as bodily resources or 'donors' of reproductive tissue. There is also an open question about who is most likely to benefit from these therapies, whether continuing health inequalities might be reduced, or whether these technologies will be distributed equitably. Debates surrounding the risks and benefits of ova donation for therapy and for research highlighted complex ways in which the global bioeconomy might benefit some women at the expense of others. These complexities are a focus for international attempts to eradicate trafficking, improve safety and set standards for good science and good clinical practice. Even here, however, disparities and difficulties were highlighted, as the lack of international consensus about ethical concerns, diverse cultural values associated with human tissues, and competition in an international marketplace shaped policy and practices.

Internationally, regulation is seen as weak in controlling commercial exploitation of unproven 'treatments'. Legal frameworks for regulating human-tissue use, even in the British context, despite claims to the contrary, could be seen as unable adequately to specify or fix legal subjectivity, or the meaning of embodiment or personhood. Abortion law in Britain was characterized as paternalistic and failing to support rights-based claims to abortion care, while legal attempts to distinguish between persons and things (in patent law) or between human and non-human entities could be seen as increasingly fraught and contested. In short, while political support for tissue- and cell-based therapies in Britain and elsewhere has been strong, realizing the potential claimed for them depends on evidence that they are effective and on a much wider social and economic commitment to transform innovation stories into the delivery of tangible health benefits for all.

Appendix

Table A.1 Chronology of developments in UK legislation and policy related to human-tissue and stem cell research

1984	Report of the Committee of Inquiry into Human Fertilisation and Embryology (Cm 9314), 1984. The Warnock report.
1987	*Human Fertilisation and Embryology: A Framework for Legislation* (Cm 259). This White Paper, which was published in November 1987, preceded the HFE Act and indicated government commitment to legislation in this area.
1989	*Polkinghorne Guidelines: Review of the Guidance on the Research Use of Fetuses and Fetal Material* (Cm 762).
1990	Human Fertilisation and Embryology Act 1990. The Act provided for regulation of the creation or use of embryos outside the body; the use of donated eggs or sperm in treatment; and the storage of embryos, sperm or eggs.
2000	*Stem Cell Research: Medical Progress with Responsibility.* Sir Liam Donaldson chaired the report of this expert advisory group, commissioned to assess the anticipated benefits of research on SCs and cell nuclear replacement and to advise whether further research uses of embryos should be permitted. The report also recommended that, 'mixing of human adult (somatic) cells with the eggs of any animal species should not be permitted'.
2001	The Human Fertilisation and Embryology (Research Purposes) Regulations 2001. Extended the purposes for which research licences could be authorized, namely: increasing knowledge about the development of embryos; increasing knowledge about serious disease; or enabling any such knowledge to be applied in developing treatments for serious disease.
2001	The Human Reproductive Cloning Act 2001. This Act created an offence of placing a human embryo in a woman other than created by fertilization.
2002	House of Lords Select Committee report on Stem Cell Research (2002). In direct reference to human embryos only, the Committee believed that, 'embryos should not be created specifically for research purposes unless there is a demonstrable and exceptional need which cannot be met by the use of surplus embryos'. The Committee also discussed the issue of animal–human hybrid and chimera embryos and took issue with the suggestion of Sir Liam Donaldson's expert group that there is a need for an outright ban on such research.
2004	Human Tissue Act 2004, which established the Human Tissue Authority in 2005.
2005	House of Commons Science and Technology Select Committee report on Human Reproductive Technologies and the Law, 2005.
2005	UK Stem Cell Initiative. Report and Recommendations. The UK Stem Cell Initiative was chaired by Sir John Pattison and reported in November 2005.

In response to the report's recommendation that SC research needed sustained and increased funding, the government increased its funding over the two-year period 2006–7 to 2007–8 from £50 million to £100 million.

2005   Government Consultation on the Review of the Human Fertilisation and Embryology Act. Launched in August 2005, the consultation posed a wide range of questions about how the law might be updated.

2006   Applications to HFEA for licences to derive SCs from human embryos, created from animal eggs instead of human eggs, November.

2006   Government produced a White Paper, *Review of the Human Fertilisation and Embryology Act: Proposals for revised legislation (including establishment of the Regulatory Authority for Tissue and Embryos)* (Cm 6989) that included, among other things, a proposal to establish Regulatory Authority for Tissue and Embryos (RATE) to replace the Human Tissue Authority and the Human Fertilisation and Embryology Authority (HFEA), December.

2007   House of Commons Select Committee on Science and Technology announced inquiry into the regulation of hybrid and chimera embryos, January.

2007   Academy of Medical Sciences initiated a study to examine the issues surrounding the creation of interspecies embryos, February.

2007   Science and Technology Committee published its report on the 'Government proposals for the regulation of hybrid and chimera embryos', March.

2007   Human Fertilisation and Embryology Authority undertook a public consultation on 'Hybrids and chimeras', April.

2007   House of Lords stem cell debate, May.

2007   Draft Human Tissue and Embryology Bill published for pre-legislative scrutiny, May.

2007   Joint Parliamentary Scrutiny Committee appointed, May.

2007   The Academy of Medical Sciences published its report on 'Inter-species embryos', June.

2007   The Joint Parliamentary Scrutiny Committee published its report. Many of the key recommendations the Trust had made, including support for interspecies embryos and not introducing proposals for RATE, were supported in the report, August.

2007   HFEA published its report on its consultation on 'Hybrids and chimeras', September.

2007   Government published its response to the Joint Committee Report, October.

2008   Human Fertilisation and Embryology Bill introduced in the House of Lords. The Bill as introduced does not seek to amend the Human Tissue Act (2007), nor does it propose to introduce RATE, the Government having dropped the latter proposal following widespread objection from the communities it would regulate and the Parliamentary Scrutiny Committee, December.

2008   Bill debated in the House of Lords – the Trust, with the Medical Research Council, Academy of Medical Sciences and Royal Society, provided briefings to peers at all stages of the debate, January.

2008   HFEA issues licences for use of human–animal cytoplasmic hybrid embryos at King's College London and Newcastle University, January.

2008   Human Fertilisation and Embryology Bill passed in the House of Lords and introduced in the House of Commons, February.

2008   Human Fertilisation and Embryology Bill receives Royal Assent, November.

Source: Adapted from House of Commons STC Fifth Report 2007 on Hybrids and embryos and the Wellcome Trust History of the HFE Bill, available at: www.wellcome.ac.uk/About-us/Policy/Spotlight-issues/Human-Fertilisation-and-Embryology-Act/History/. See also: www.hfea.gov.uk/ and www.hta.gov.uk/ for details of press releases and policy developments in this area (last accessed April 2011).

Notes

1 Commodifying tissues and cells: the new tissue economies

1 See www.fda.gov/NewsEvents/Newsroom/PressAnnouncements/2006/ucm108589. htm

2 HTA Regulatory alert 001/2006 requested that all establishments storing material on behalf of ACT/Biomark/CellTech/BioCells or that originated from ACT/Biomark/CellTech/BioCells contact the authority and hold the cells for safety testing and investigation.

3 Email from ACT customer support dated 10 September 2007 confirmed that patients were no longer being treated on a commercial basis. In September 2010, Dr Robert Trossel, who had links with ACT/Biomark was struck off the General Medical Council's register of medical practitioners for misconduct and for misleading patients from the United Kingdom, treating them by injecting them with SCs unsuitable for use in humans; see www.bbc.co.uk/news/health-11439711 (last accessed January 2011). See also Blackburn-Starza (2010).

4 In contrast, Rajan argues that, 'the life sciences are *overdetermined* by the capitalist economic structures within which they emerge' (Rajan 2006:6). However, he and Rose seem to share the view that capitalism is not a unitary system that necessarily determines the direction of the development of the life sciences, but rather there are multiple forms of it and multiple possibilities.

5 See, for example, the UK National Health Service Blood and Transplant Service (NHSBT). The NBTS works with these 'traditional tissues' and improving tissue-processing techniques; in its research and development programme, it is also working to develop innovative tissue-engineered therapies 'which can grow with the patient'; see www.nhsbt.nhs.uk/ (last accessed January 2011).

6 In using this term, Lock deliberately draws attention to the blurred boundary between life and death and highlights the ambiguous status, or 'hybridity', of the body at this time.

7 For other anthropological accounts of the procurement of organs, see Fox (1992) and Lock (2000, 2002b).

8 So, for example, Hogle describes how a transplant co-ordinator judges that certain types of person make good donors, not simply because of their health status or the condition of their body or organs, but rather because their social circumstances make donation a good thing, perhaps by giving meaning to an otherwise worthless and meaningless life (Hogle 1999).

9 Lock (2002b) says that, in Japan, the notion of gifts as reciprocal exchanges is important, and it is for this reason that giving an organ to a complete stranger seems

odd to many Japanese. She concludes, therefore, that the concept of altruism is not well developed there (p10).

10 See http://handsoffourovaries.com/, an international campaign group that objects to the collection of ova for research (last accessed 26 January 2011).

11 See www.transeuro.org.uk/index.html (last accessed 26 January 2011).

12 As one interviewee, a UK scientist (S5), explained, 3T3 mouse fibroblasts are the commonly used feeder layer for keratinocytes, but they were researching the use of human fibroblasts as feeder cells, with the aim of developing serum and animal-free products. 3T3 is the industry-standard cell line, originally derived by Green *et al.* in the United States. Controversy in the United Kingdom centred around attempts by the UK Xenotransplant Interim Regulatory Authority to seek assurances from researchers using these in skin products that no viruses were present. Researchers were expected to be able to give assurance that as yet unidentified viruses could not be a risk.

13 Lysaght and Crager trace the origins of 'tissue engineering' and the first use of the term to 1982 and the work of Eugene Bell, who later founded Organogenesis (Lysaght and Crager 2009).

14 See work by Molly Stevens *et al.* (2005) at Imperial College and spin-out company Bioceramic Therapeutics at www.bioceramictherapeutics.com/index.html (last accessed January 2008) and www.repregen.com/ (last accessed 31 January 2011).

15 It has also been suggested that menstrual blood is a source of SCs, and a company in the United States, C'elle, has begun to offer women the opportunity to store their menstrual blood; see www.celle.com (last accessed January 2008).

16 According to Ang, the practice of nose mutilation or amputation as a means of humiliating those charged with various social crimes led to reconstructive surgery where a flap of skin is excised from one part of the body (the face in this case) and relocated, while still being attached to the original site in order to maintain a blood supply to the tissue. It is still used in some plastic-surgical procedures today. Free skin grafts are where the skin is detached and then transplanted to a new site. For a cultural history of aesthetic surgery, see Gilman (1999).

17 See www.euroskinbank.org (last accessed November 2007). See also Huang *et al.* (2004) for a discussion of this method in the UK National Blood Service tissue-banking service, which provides both types of allograft skin – cryopreserved living skin and glycerol-preserved non-living skin.

18 For a discussion of the use of allogeneic skin in burns treatment, see, for example, Burd and Chiu (2005).

19 Two earlier ones were set up at the University Hospital in Hradec Kralove, in 1952, and a Yorkshire Regional Tissue Bank in the United Kingdom (Pianigiani *et al.* 2005, Kearney 2006).

20 For a fuller description of quality-control issues in the skin bank, see, for example, Alotto *et al.* (2002).

21 See www.euroskinbank.nl/ (last accessed 31 January 2011).

22 Orginally developed by BioSurface Technology, in 1987, and acquired by Genzyme, in 1994, together with an autologous cartilage product Carticel, launched in 1995. See www.genzyme.com for further details and note especially the culturing of the cells brings them into contact with animal materials: Epicel should not be used in patients with known sensitivities to materials of bovine origin. The cell-culture medium used in the culture of Epicel contains bovine serum. The medium used to package and transport Epicel does not contain serum; however, trace

quantities of bovine-derived proteins may be present. This tissue is intended for autologous use and has not been tested for biohazards. Health-care providers should handle this product as if infectious agents are present. During the Epicel manufacturing, patients' cells are co-cultured with mouse cells. Although the mouse cells have been tested and found to be free of bacteria, fungi and virus, an infection cannot be excluded. As a safety measure, the Epicel-treated patients are precluded from donation of blood or blood parts, tissue, breast milk, egg, sperm or other body parts for use in humans (last accessed November 2007).

23 In one description of the therapy, Epicel is combined with the use of temporary allografts and other complex wound-management techniques (Carsin *et al.* 2000). It takes approximately sixteen days to culture the skin, which then has a 'shelf life' of only twenty-four hours. Therefore, the patient has to be sustained while the skin is prepared.

24 See Zhu *et al.* (2005) and Moustafa *et al.* (2004) for a discussion of the delivery method and growing cells directly on the surface without bovine collagen or murine feeder cells, treatment of burns, chronic wounds and diabetic foot ulcers. Another method of delivering autologous keratinocytes being tested in the United Kingdom is to spray them on to a skin substitute, Integra; see www.blondmcindoe.com (last accessed November 2007). See Kearney (2001) for a fuller discussion of skin substitutes and 'composite grafts', which combine an allogeneic or autologous skin graft with synthetic matrices.

25 In 2006, each application of Myskin cost the purchaser around £120.

26 The use of neonatal foreskin appears to have been relatively uncontroversial, although one interviewee, in 2003 (S7), described an encounter with a UK pressure group NORM, which opposes circumcision and the use of foreskin in such products; see www.norm-uk.org (last accessed November 2007).

27 There is an extensive clinical literature on the use of Apligraf; see www. organogenesis.com for details (last accessed November 2007).

28 In an interview in 2004, I was told that the relationship between Novartis and Organogenesis broke down, leading to Novartis withdrawing financial support and application for approval of Apligraf in Europe being withdrawn. Apligraf has, to date, only been available in the US market. In our interviews in 2003, shortly after the collapse of ATS and Organogenesis, respondents discussed the absence of a successful business model for tissue-engineered products and the need for new business models, suggesting that these failures were because there were aspects of the pharmaceutical model that didn't fit with these types of product. In particular, the lack of large-volume sales, but with high start-up costs and long lead times, was seen as problematic. In addition, the regulatory hurdles were seen as hampering the process of getting products to market. See also Naughton (2001) and Williams (2002).

29 The Naughtons filed an objection to the sale of Dermagraft to S&N, saying they had been granted rights to have the patents returned to them if ATS failed, but the courts found the claim unsupported in 2003. Gail Naughton became scientific advisor to Theregen, a company developing a treatment for ischaemic heart disease using Dermagraft as a patch for damaged heart tissue. The patch is believed to encourage blood circulation to the heart. See www.theregeninc.com (last accessed November 2007) and Siani-Rose (2006). For discussions of Dermagraft in the clinical literature, see, for example, Krishnamoorthy *et al.* (2003), Omar *et al.* (2004), on use in chronic venous leg ulcers, and Marston *et al.* (2003) on use in diabetic foot ulcers.

30 See also Lysaght (2004).
31 Presentation by Geoff McKay to European Stem Cells Regenerative Medicine Congress, May 2007, London.
32 See www.intercytex.com (last accessed November 2007).
33 Interview with Smith & Nephew employee, 2003. Another interviewee discussed the importance of randomized controlled trial data to support the use of products such as Dermagraft, arguing that such data, together with the regulatory controls applied to drugs, were needed. Such a view adopts a classical position on the need for certain types of clinical evidence to demonstrate efficacy.
34 One estimate of burns patients requiring tissue-engineered skin grafts in Western Europe is 150 per year (Husing *et al.* 2003). For a discussion of the role of skin substitutes and some engineered skin products in burns treatment, see, for example, Shakespeare (2005).
35 Personal communication from Lee Laurent-Applegate, 14 November 2007.
36 Harding *et al.* argue that there is a lack of clinical support for developing high-quality wound-care services and research into the value of new approaches to wound management (Harding *et al.* 2002).
37 Allografts, however, are less likely to be understood as intimately tying the recipient and donor together in ways that heart, liver and kidney donors have described, but instead are shaded by a rather different psychology (Panconesi and Andreassi 2005).

2   Regenerative medicine: a paradigm shift?

1 See www.remedi.ie/ (last accessed February 2008).
2 An example of this was set up in 2007, Stem Cells for Safer Medicines (SC4SM); see www.sc4sm.org (last accessed 1 April 2011).
3 See www.millipore.com/press/pr3/rencell; Millipore (NYSE: MIL) is an international life-science company providing cutting-edge technologies, tools and services for bioscience research and biopharmaceutical manufacturing. The ReNcell neural SC lines are immortalized human neural progenitor cells that have the ability readily to differentiate into neural cell types and will replicate indefinitely. ReNcell immortalized cells display the same marker patterns as normal cells and are able to grow and remain stable after culturing. These cells are available as tools for the discovery of new therapies targeting diseases of the CNS. Interestingly Millipore also distributed two human embryonic SC lines from Australia: MEL-1 and MEL-2. These had been accepted by the UKSCB but not yet accessioned (February 2007).
4 Food and Drug Administration 21CFR Part 1271 Human Cells, Tissues, Cellular and Tissue-Based Products, AATB 'Standards for Tissue Banking' proposed changes, 2006. The United States restricts the use of tissue for transplantation or therapies from countries at risk of vCJD.
5 See www.reneuron.com/ (last accessed May 2007); see also Annual Report 2006. There are an estimated 50 million stroke survivors worldwide, approximately one half of whom are left with permanent disabilities. The annual health and social costs of caring for these patients are estimated to be in excess of £5 billion in the United Kingdom and in excess of US$50 billion in the United States.
6 It had attracted investment from Merlin Biosciences and grants from the European Research Framework, UK DTI and Michael J Fox Foundation. In 2006, it was

floated on the Alternative Investment Market, on the London Stock Exchange, but made an overall loss for the year of £6.3 million (ReNeuron 2006).

7 See www.clinicaltrials.gov, Clinical Trial NCT01151124 Pilot investigation of stem cells in stroke (PISCES) (last accessed February 2011).

8 See Alison Kraft fellowship description at www.york.ac.uk/res/sci/fellowships/res350270005kraft.htm (last accessed 1 April 2011).

9 See previous note.

10 See www.ucdenver.edu/academics/colleges/medicalschool/departments/medicine/ClinicalPharmacologyToxicology/Pages/CurtFreed.aspx (last accessed 24 April 2007).

11 MRC project grant (N.D. Allen, A.E. Rosser and S.B. Dunnett) G0701489: Development of a platform to generate clinical-grade neural progenitors for transplantation in Huntington's disease; £654,348 over three years, March 2008.

12 Prior to techniques of hibernation (a method of storing the cells by refrigeration at above freezing; see www.patentstorm.us/patents/6713245-description.html), the recipient patient would have to be prepared for surgery at the same time that the aborted tissue was being collected.

13 Because of abortion methods, in other senses, fetal tissue has been regarded as immunologically privileged, uncontaminated and therefore especially valuable as a research tool (Hall 2003).

14 Medical abortions are induced by the administration of drugs, and, because the fetus is passed by the women, it is more likely to be in tact and easier to dissect than a fetus that is aspirated during a surgical abortion. Medical abortion is also becoming more common and is a recommended method of terminating early-stage pregnancy (Royal College of Obstetricians and Gynaecologists 2004).

15 Anders Bjorklund presentation to the 18th Annual Meeting of NECTAR, Lund, Sweden December 2007. See www.transeuro.org.uk (last accessed February 2011).

16 For a discussion of the emergence of UK Science Plc, see Jasanoff (2005:chp9).

17 This line was derived from the lungs of a 16-week female fetus in 1977, collected following the termination of pregnancy of a thirty-eight-year-old white woman, mother of six (Leiva 2006, Yu *et al.* 2007).

3 Regulation and governance of tissue- and cell-based therapies in Europe: ethical controversy and the politics of risk

1 See http://archive.eurostemcell.org/Documents/Ethics/W06/Workshop06Report. pdf: Ethical aspects of commercialization and patentability of stem cells and stem cell lines (last accessed 21 February 2011).

2 Waldby and Salter (2008) also discuss the effects of the TRIPS and WTO patent requirements, which goes beyond my discussion here but has implications for SC research internationally.

3 European Directive 2004/23/EC on setting safety standards on quality and safety for the donation, procurement, testing, processing, preservation, storage and distribution of human tissues and cells (the Tissues and Cells Directive) (Official Journal L 102 7/4/2004 p48 -58); Regulation (EC) No. 1394/2007 (Official Journal L 324, 10 December 2007, pp. 121–37).

4 Council of Europe, Recommendation No. R (94) 1(Council of Europe 1994).

5 Tatarenko (2006) describes a 'close co-operation' between the Council of Europe and EC Commission and how the Council of Europe first identified concerns about

tissue safety in the 1970s and developed its Guidance on safety of tissues (Council of Europe 2002).

6 The interviews were carried out as part of an ESRC-funded project on the regulation and governance of tissue-engineered products in Europe and UK Grant L218252058. The fieldwork was conducted by Alex Faulkner, Ingrid Geesink and myself.

7 This same point was made by Peter Liese in the Debate of the EU Parliament on the Second Reading on 15 December 2003:

> We have also made clear that there is to be no trading in cells and tissues as such, but that there is no objection to trading in medicines manufactured from them. As is made clear in one of the recitals, industry has an important part to play in this.

And Byrne, for the Commission, said,

> The Council of Europe and the European Group on Ethics have stressed the principle of not-for-profit procurement of tissues and cells. The directive should therefore accord with this principle. However I must draw a distinction between the act of procurement, to which the 'not-for-profit' principle should apply, and the secondary activities, such as the further processing, manipulation, testing or the manufacturing of products, to which it should not apply.

8 See www.europabio.org/; EuropoBio and Eucomed produced a number of position papers on the drafts of the Tissue and Cells Directive and the Council Common Position; see, for example, Joint EuropaBio/Eucomed Position Paper to the Report from Dr Lieses for the Proposal for a Directive of the EUP and of the Council on setting standards of quality and safety for the donation, procurement, testing, processing storage and distribution of human tissues and cells (EC Brussels, 19 June 2002; COM 2002 319 final); EuropaBio and Eucomed (revised) comments on Common Position by the Council (14 July 2003) of the Sanco Directive on 'Setting standards of quality and safety for the donation, procurement, testing, processing, storage, and distribution of human tissues and cells', 2002/0128 (COD).

9 CARE Evidence Statement by David Fieldsend, Office Manager, Brussels, on the Proposed EU Human Tissues Directive Public Hearing, January 29 2003.

10 Directive 2004/23/EC of the European Parliament (2004) and of the Council as regards traceability requirements, notification of serious adverse reactions and events and certain technical requirements for the coding, processing, preservation, storage and distribution of human tissues and cells, Official Journal L294 25/10/2006 pp32–46; Directive 2004/23/EC of the European Parliament and of the Council as regards certain technical requirements for the donation, procurement and testing of human tissues and cells, Official Journal L38 9/2/2006 pp40–52.

11 Article 3 (o) 'tissue establishment' means a tissue bank or a unit of a hospital or another body where activities of processing, preservation, storage or distribution of human tissues and cells are undertaken. It may also be responsible for procurement or testing of tissues and cells.

12 Since then, a new EU Directive on standards of quality and safeaty of human organs intended for transplantation has come into effect: Directive 2010/45/EU Official Journal L207/14 6.8.2010.

13 Commission Directive 2003/63/EC 23 June amending Directive 2001/83/EC in Offical Journal L159 27.6.2003.

14 EU Parliament Committee on the Environment, Public Health and Safety Draft Report Provisional 2005/0227 (COD).

15 MEPs set out safer rules for innovative therapies. See www.euractiv.com/science/ eu-leaves-decision-stem-cell-ethics-member-states/article-163378 (last accessed 10 May 2007). These three groups are PES, The Socialist Group, ALDE, Alliance of Liberals and Democrats, and GUE/NGL, European United Left and Nordic Green Left. The rapporteur claimed that he and his report represented the majority view of the Committee.

4 A 'strict but permissive approach": a case study of UK regulation of human-tissue and cell therapies

1 For example, see www.talkingstemcells.ed.ac.uk/ for details of the public-engagement project funded by the UK research councils ESRC and BBSRC. For public-engagement activities at North East England Stem Cell Institute, see www.nesci.ac.uk/research/discussion/. And, on 6 October 2010, there was an International Stem Cell Awareness Day; see www.stemcellday.com/ and also www. youtube.com/UCLTV?feature=mhum#p/u/12/k0-XoqMh6Jk (all last accessed 11 March 2011).

2 Feminist International Network of Resistance to Reproductive and Genetic Engineering (FINNRAGE) opposed these technologies.

3 See, for example, SPUC Human Fertilisation and Embryology Bill Commons Second Reading, 12 May 2008, summary briefing for MPs; press releases by LIFE between 5 July 2007 and 21 May 2008, when LIFE described the result of the parliamentary vote to reject amendments to the abortion time limit as 'bad news'; see www.spuc.org.uk/ and www.lifecharity.org.uk/ (last accessed May 2008).

4 See www.progress.org.uk (last accessed May 2008). Subsequently, in 1992, it became a charity: PROGRESS Educational Trust (PET).

5 Lord Winston's support for the HFEA, however, seemed to waiver, and he was subsequently quoted as saying that it should be abolished and replaced, BBC News, Friday 10 December 2004: http://news.bbc.co.uk/1/hi/health/4084365.stm (last accessed May 2008).

6 See www.legislation.gov.uk/ukpga/1967/87/section/1. The Abortion Act 1967 was amended by the 1990 HFE Act to reduce the gestational limit to 24 weeks. For a discussion of the abortion law debate, see www.prochoiceforum.org.uk/ index.php (last accessed March 2011).

7 For a legal account of the HFEA Act and its effects, see Morgan and Lee (1991).

8 Holding a Progress Educational Trust event, 'Half-truths?: the science, politics and morality of hybrid embryos', which took place at the Houses of Parliament on the evening of Monday 19 May 2008. It also convened a conference, 'Debating deafness and embryo selection: are we undermining reproductive confidence in the deaf community?', in Cardiff, in April 2008, to debate clause 14(4)(a), and another, 'Reproducing regulation: new laws for fertility treatment and embryo research – will we get it right?', at the Institute of Child Health, London, in November 2007.

9 See www.hfea.gov.uk (last accessed March 2011).

10 See www.corethics.org/ (last accessed March 2011).

11 Jasanoff suggests that the HFEA consultation opened the way for SPUC to fund its private campaign to influence public opinion, and therefore this public-relations strategy was seen by some as anti-democratic (Jasanoff 2005:157).

12 Jasanoff (2005:58).
13 Since the 1960s, when Harris and Watkins first fused human and mouse cells (paper presented to EGENIS workshop, July 2008, by Duncan Wilson).
14 House of Commons Science and Technology Committee (2005). Half of the Select Committee disagreed with this report and published its own special report: HoC Science and Technology Committee Inquiry into New Reproductive Technologies and the Law, Eighth Special Report, March 2005; Review of the Human Fertilisation and Embryology Act, A Public Consultation, Department of Health, 2005.
15 See minutes of HFEA meetings, November 2007 and January 2008, www.hfea. gov.uk (last accessed May 2008).
16 It was encouraged to adopt a policy in favour of mixing human and animal cells by Dr Evan Harris, a Liberal Democrat MP and member of the Science and Technology Committee, and a group of leading scientists, in a letter to *The Times* in January 2007.
17 Retrospective on the Human Fertilisation and Embryology Act 2008, convened by Innogen at 1 Birdcage Walk, March 2009.
18 See, for example, press releases from the Science Media Centre, 12 May 2008 and 19 May 2008.
19 The Ethics Committee and the Scientific Advisory Committee of the RCOG, in response to the proposed revisions to the legislation, argued that the HFE Act and the HFE Authority represented a highly successful and robust model for regulating assisted reproduction, widely admired throughout the world and successful in maintaining a balance of concern between the interests of patients, professionals, scientists and the public at large in a controversial area, and that a merger with the HTA – 'an unproven body' – would create a remit that was 'too wide' to be effective (Royal College of Obstetricians and Gynaecologists 2005a 2005b).
20 See www.practicalethicsnews.com/practicalethics/2008/05/hfea-and-regula.html (published 22 May 2008; last accessed June 2008).
21 In her comparison of the British approach to regulating embryo research in the 1990s with that of Germany and the United States, Jasanoff suggests that, in Britain, the rights of women were hardly discussed at all; that Britain was more permissive, but handled novelty on a case-by-case basis under centralized regulatory control; that a dominant coalition of scientists and the state succeeded in speaking for the embryo; and the HFEA guarded against the slide down a 'slippery slope' into moral anarchy or disorder (Jasanoff 2005:170).
22 Personal communication from HFEA, 26 June 2008.
23 The NHS Retained Organs Commission was set up to advise hospitals and families of the disposal of 'retained organ's, until its closure in March 2004.
24 UKXIRA was disbanded in 2006, a move that attracted some criticism (McLean and Williamson 2007).
25 HTA statement on RATE White Paper, 14 December 2006; see www.hta.gov.uk (last accessed June 2008).
26 Human Tissue Authority briefing on the implications of the Joint Committee Report on the Human Tissue and Embryos (Draft) Bill, 31 October 2007; see www. hta.gov.uk (last accessed June 2008).
27 Most, but not all, countries have a single competent authority; see Summary Table of Responses from Competent Authorities: Questionnaire on the transposition and implementation of the European Tissues and Cells regulatory framework, SANCO

C6 CT/gcs D (2007) 360045, Brussels, 6 February 2007, which noted that the UK HFEA and HTA provided two separate reports.

28 HTA Position Statement on regulating human embryonic stem cell lines for human application. Joint statement from the HTA, HFEA and MHRA. Originally issued May 2007; updated May 2008; see www.hta.gov.uk (last accessed June 2008).

29 See Interim UK Regulatory Route Map for Stem Cell Research and Manufacture at www.mhra.gov.uk/Howweregulate/Medicines/Medicinesregulatorynews/ CON041337 and Tool Kits at www.sc-toolkit.ac.uk/home.cfm (last accessed March 2011).

30 Seven of the fourteen members of the committee who produced or contributed to the Code were from industry.

31 The Global Harmonisation Taskforce group was also discussing tissue products at this time.

32 For an explanation of the role of Notified Bodies in device regulation, see Kent *et al.* (2002b) and www.mhra.gov.uk/Howweregulate/Devices/NotifiedBodies/ index.htm (last accessed March 2011).

33 It was disbanded in 2003 and superseded by the Scientific Committee for Emerging and Newly Identified Health Risks (SCENHIR) in 2004.

34 Policy statement on the relationship between the RATMP and the Quality and Safety Regulations Joint Statement from the HTA and the MHRA; 19 March 2008; see www.hta.gov.uk (last accessed June 2008).

35 See www.reneuron.com/monthly-archive/159-reneuron-gains-uk-regulatory-approval-to-start-ground-breaking-clinical-trial-with-stem-cell-therapy-for-stroke—190109 for company press releases announcing approval of the trial to commence in the United Kingdom (last accessed June 2009).

36 A narrower remit than first appears, as other SC trials will be reviewed by NHS Research Ethics Committees; see www.advisorybodies.doh.gov.uk/genetics/gtac/ index.htm (last accessed July 2008).

37 Research Governance Framework for Health and Social Care, February 2001. Revised edition, April 2005.

38 See Note 29.

5 'Football fields of skin": a masculinist dream?

1 Personal communication from Stefan Kaelin, Organogenesis Managing Director, Switzerland 18 August 2008.

2 This was the first grantholders' meeting for those funded by the UK Joint Council Stem Cell Initiative.

3 See, for example, the UK Women's Engineering Society: www.wes.org.uk; UK Resource Centre for Women in Science, Engineering and Technology: www.theukrc.org/; WiTEC – the European Association for Women in Science, Engineering and Technology (SET): www.witec-eu.net.

4 Originating from a wealthy, aristocratic family, Anne McClaren was educated at Cambridge, had a distinguished career in genetic and reproductive science and was influential in policy circles at home in the United Kingdom and internationally – for example, as a member of the Warnock Committee and the European Group on Ethics (EGE) – until her tragic death in 2007. Alison Murdoch, a clinician by training, has been at the centre of controversies in the United Kingdom surrounding the use of eggs and embryos for SC research and cloning techniques. She is

Professor of Reproductive Medicine and Head of NHS Newcastle Fertility Centre and Department of Reproductive Medicine at University of Newcastle, a member of the UK Nuffield Council on Bioethics and Chair of the British Fertility Society.

5 For a discussion of women's careers in science in the United Kingdom, see papers from the ATHENA project: www.athenaproject.org.uk; for data on women in science across Europe, see Women and Science, Statistics and Indicators, SHE Figures 2006, 2009, European Commission DG Research at http://ec.europa.eu/research/science-society/document_library/pdf_06/she_figures_2009_en.pdf (last accessed March 2011).

6 Transplant Activity in the UK 2006–7. NHS Blood and Transplant: www.nhsbt.nhs.uk/ (last accessed August 2008).

7 See the Organ Procurement and Transplantation Network (OPTN) data on transplants performed in the United States since 1988: www.optn.org/ (last accessed August 2008).

8 US Alzheimer's Association Statistics, 2008: Alzheimer's disease facts and figures: www.alz.org/alzheimers_disease_facts_figures.asp (last accessed August 2008).

9 National Spinal Cord Injury Statistical Center, Birmingham, Alabama. Spinal-cord injury facts and figures at a glance, 2008: www.spinalcord.uab.edu/show.asp?durki=116979&site=1021&return=19775 (last accessed August 2008); 70–80 per cent of spinal-cord injury patients in the United Staes are men.

10 UK National Diabetes Audit: key findings about the quality of care for people with diabetes in England, incorporating registrations from Wales. Report for the audit period 2004–5. The Information Centre 2006.

11 For international data on 'gender and health', see, for example, World Health Organisation: www.who.int/gender/en/ (last accesssed March 2011).

12 See, for example, Patient and GP information sheet on ACI, the Grovesnor Nuffied Hospital, Chester Knee Clinic, where patients are enrolled in the Stanmore clinical trial; see www.kneeclinic.info (last accessed August 2008).

13 See also Genzyme's information for patients on Carticel (USA): www.carticel.com/; Verigen's (Germany) MACI collagen matrix technology for seeding on and growing cartilage cells (taken over by Genzyme in 2005): www.regenerationnet.com/node/151; Tigenix (Belgium) www.tigenix.com (last accessed August 2008).

14 Hyalograft C is a biocompatible scaffold system comprising hyaluron-based bio-materials . The company at its 'TissueTech' laboratories also provided autolous skin graft product 'Laserskin'.

15 See www.stepsproject.com (last accessed August 2008).

16 STEPS Executive Summary, 2nd Annual Report pXVI.

17 STEPS Executive Summary, 2nd Annual Report 2007 pV.

18 See http://handsoffourovaries.com/ (last accessed August 2008).

19 Indeed, this is where she takes issue with the commodification thesis (Cooper 2008: 190 note 20).

20 Almost every year, the number of pregnancies legally terminated in England and Wales has increased from just over 54,000 in 1969 to 193,700 in 2006. At least one in every three women will have undergone a termination by the time they reach the age of forty-five years. Abortion is carried out mainly in approved, independent clinics, before 10 weeks gestation. Medical abortions, which are recommended for terminations in early pregnancy, are on the increase – rising from 24 per cent in 2005 to 30 per cent of the total in 2006. Abortion is the most common surgical procedure on women of reproductive age. Most (87 per cent) are funded by the National Health Service (Kent 2008a, Department of Health 2007c).

21 See www.millipore.com/press/pr3/rencell
22 See www.cscr.cam.ac.uk/about-us/
23 Presentations to European Stem Cells Regenerative Medicine Congress, by Allsopp and Ilyne, London 2007.
24 See www.stemcellsciences.com (last accessed August 2008).
25 See www.computescotland.com/edinburgh-university-stem-cell-spin-out-sold-to-us-2123.php (last accessed March 2011).
26 See www.stemcellsinc.com/ (last accessed March 2011).
27 See www.clinicaltrials.gov/ct/show/NCT00337636?order=4 (last accessed August 2008).
28 See www.neuralstem.com/ (last accessed August 2008).
29 See press release, Thursday 28 June: StemCells Agrees to Stay in Lawsuit Against Neuralstem Over Disputed Patents: see www.neuralstem.com/ (last accessed August 2008).
30 See www.stemcellinnovations.com/ (last accessed August 2008).

6 Remaking the self

1 Kolata, G., 2007 *New York Times*, '*Man who helped start stem cell war may end it*', 22 November. Interestingly, what was less well reported was that one of the sources of the skin cells was from a fetal line IMR-90 (see Chapter 2).
2 Preformationist theories also can be seen as prioritizing male agency and denying active maternal involvement (Shildrick 1997:34).
3 A blastocyst is the term given to an embryo up to 14 days post-fertilization. It is illegal in Britain, under the terms of the Human Fertilisation and Embryology Act 1990 (revised in 2008), for an embryo created *in vitro* to be allowed to develop beyond 14 days.

> After about five days, the embryo consists of a ball of 50 to 100 cells called a blastocyst. It is about the size of one of the full stops on this page. The outer layer of the blastocyst forms the placenta and the inner part is made up of the 'inner cell mass', which can give rise to embryonic stem cells. At this stage the embryonic stem cells are pluripotent, meaning they are able to specialize into all the different types of tissue needed to form the human body. Embryonic stem cells can renew themselves almost indefinitely, so once they are grown in culture, scientists can work with the same embryonic stem cells repeatedly.
>
> (Medical Research Council: Stem Cell, issued April 2008; www.mrc.ac.uk document MRC003374; last accessed July 2009)

4 See Chapter 3 for a brief discussion of the dispute relating to the patentability of embryonic SCs.
5 See www.reneuron.com. ReNeuron recruits first patient to PISCES clinical trial, press release 16 November 2010. In 2010, a new project led by Dr Roger Barker was funded by the EU FP7 to recommence trials of fetal-cell transplants in Parkinson's patients; see www.transeuro.org.uk/ (last accessed March 2011).
6 For a discussion of how reproductive rights draw on liberalist ideas of subjects, see Kent (2000) and Bordo (1995).
7 And as amended in the 1990 Human Fertilisation Embryology Act.
8 As Lee and Jackson also point out, 'a woman must frame her reasons for wanting to terminate her unwanted pregnancy in terms of its risk to her health, and must

submit to the scrutiny of medical experts before she can obtain a legal abortion' (Lee and Jackson 2002:121).

9 There has been extensive debate about the strategic potential of a rights-based approach to reproductive freedom and abortion; see, for example, Smyth (2002).

10 Ariss was commenting on the proposal to harvest ova from aborted fetuses, as discussed by the UK HEFA in 1994 (Human Fertilisation and Embryology Authority 1994).

11 Young (1984) offers an early account of these issues, especially problematizing the maternal subject and arguing that pregnancy represents a paradigmatic case of 'split subjectivity'. Also, 'The capacity to be self and other in pregnancy, which is the potential of every woman, is the paradigm case of breached boundaries' (Shildrick 1997:35).

12 See also Pfeffer (2008) for a discussion of findings from these focus groups. As the women who had donated tissue couldn't remember the detail of what they had been told about the research it would be used for, we don't assume it was used for SC research.

13 This first, pilot group comprised a group of colleagues and contacts of the researchers; the other groups were recruited with the assistance of a professional recruiter, who used her own contacts and networks to find women who had experience of abortion and some who did not. A total of forty-one women took part in the focus groups, thirty-one of whom had had an abortion. The women who had not had an abortion met separately, because it was felt that women who had an abortion would be more likely to talk freely about their experiences and views in a group that had all had abortions. All women were of reproductive age, between nineteen and forty-five years old.

14 See Pfeffer (2008): twelve participants had donated blood, and fifteen had registered as an organ donor.

15 Debates around the morality of abortion and the status of the fetus are extremely well rehearsed in a wide range of literature.

16 A distinction is made here between totipotent and pluripotent cells, which is a feature of British patent law rather than scientific agreement (Cooper 2008). The embryo/zygote produced from the sperm and ovum is a 'totipotent' cell that can produce all the cells of the body, but cells taken from the blastocyst cannot become a complete body – the embryo is destroyed by this process.

17 Derksen's study of a TE laboratory gives an interesting description of how tissue engineers try to visualize time through the use of sacrificed animals at different developmental stages; as she points out:

> the results do not show the formation of one tissue over time. Rather they tell researchers something about different tissues sacrificed at different points in time. Presumed similarity between the tissues is necessary to create an idea of tissue formation.

This rather crude method is being supplemented by the use of molecular imaging technology, which has the potential to show 'bodily processes in real time on a molecular level' (Derksen 2008:69).

18 Thanks to Mechteld-Hanna for this point and her comment that we seemed to reinforce this notion in our paper. She describes how tissue engineers see self-regulation and the immunological rejection of 'foreign cells and tissues' as an engineering problem (Derksen 2008).

19 See Chapter 3.
20 I also note that we specifically discussed the use of ACI and not other, so-called 'autologous', applications. Moreover, the use of SCNT, whereby the nucleus of a somatic cell is extracted and inserted into a denucleated egg, is said to produce 'autologous' cells.
21 For feminist accounts of the immune system, see Haraway (1995) and Martin (1994).
22 As described above in Chapter 2, Sweden was one of the first sites where fetal-tissue transplants took place, and this study included patients from the first trial.
23 A PET scan or magnetic camera can be used to locate them.

7 Life, death and immortality

1 See the Lock and Nguyen (2010) argument that such claims gloss over the changes that lead to an increased demand for organs: for example, the growing number of cases seen as operable and changes in clinical practices.
2 The first fetal neural SC line to be accepted by the UKSCB was expected to be released later in 2010 (UKSCN Bulletin, September 2010).
3 Sniff Nexoe kindly showed me around the collection and told me why it was currently hidden from public view in locked cupboards.
4 Morgan herself notes the ambiguity of both terms and the instability of distinctions between embryo and fetuses; how their meanings are not fixed, even though the scientific elucidation of stages of development appeared to make such distinctions. The 'commodification' of fetal tissue, then, following her account, also came later, and beliefs that it should be donated freely and not purchased dominated the ethos of the embryo collectors in her historical account. She also notes that 'dead fetuses' are commonly associated with induced abortion, even though many are spontaneously aborted.
5 This is especially interesting given that decisions about which embryos to implant are, in practice, more complicated than such a view suggests (Ehrich *et al.* 2007).
6 He claimed to use the techniques of SCNT to produce patient-specific SCs. This involves taking a human ovum, removing the nucleus and inserting the nucleus of a somatic cell from another person.
7 See www.isscr.org// (last accessed March 2011).
8 See also Grear (2004), who discusses the more recent Jepson case – the curate who brought a case arguing that it was unlawful to terminate a pregnancy for cleft palate. Grear's discussion points to the ways in which the 1967 Abortion Act secured medical autonomy and failed to engage adequately with the issue of the rights of either the woman or the fetus. See the Pro-choice forum for updated discussions of the lobby to reform British abortion law: www.prochoiceforum. org.uk/ (last accessed August 2010).
9 For a discussion of fetal-tissue disposal in the US context, see Ariss (2003) and Morgan (2002).
10 See, for example, collection of cord blood: Future Health, www.futurehealth.co.uk/; Smartcells, www.smartcells.com/; collection of menstrual blood: LifeCell, www.lifecellinternational.com (based in India); Celle, http://www.celle.com/ (part of the Florida-based company CryoCell, www.cryo-cell.com (last accessed September 2010); collection of placental cells: www.istemcelli.com/placentalsc. html. A new International Placenta Stem Cell Society was established in 2009,

'to promote research on all aspects related to knowledge, experimentation and clinical use of placenta-derived stem cells': www.iplass.net/ii/ (last accessed September 2010).

11 For a discussion of public cord-blood banks, see Busby (2010a, 2010b).

12 See also Bharadwaj and Glasner (2009) and Lock and Nguyen (2010) for a discussion of colonialism and the exploitation of indigenous groups in the establishment of national biobanks and bioprospecting. Also Anderson (2000) on the colonial exploitation of the brains of the *Kuru*.

13 It has been suggested that such concerns influence commercial decisions to forgo European markets and focus instead on the potentially lucrative and more accessible US market for therapies.

14 See EMEA European Medicines Agency (2009) Committee for Medicinal Products for Human Use Summary of positive opinion for Chondrocelect. EMEA/CHMP/ 383366/2009, 25 June. Tigenix is a Belgium-based company that markets this product; see www.tigenix.com/ (last accessed August 2010).

15 See www.parkinsons.org.uk/default.aspx?page=10997 and www.mssociety.org.uk/ research/az_of_ms_research/sz/stem_cells.html (last accessed March 2011).

16 Four years later, Dr Trossel was struck off by the General Medical Council for his part in promoting the work of Advanced Cell Therapeutics and referral of UK patients with MS to a Dutch clinic for treatment with SCs: BBC News, 29 September 2010; see http://news.bbc.co.uk/1/hi/health/8452874.stm.

17 See www.closerlookatstemcells.org (last accessed March 2011).

18 See www.nesci.ac.uk/news/item/egg-sharing-women-to-get-help-with-ivf-treatment-costs-for-donating-eggs-to-research (last accessed 29 October 2010).

Bibliography

Allie, D., Hebert, C., Lirtzman, M., Wyatt, C., Keller, V., Souther, S., Allie, A., Mitran, E. and Walker, C. (2004) Novel treatment strategy for leg and sternal wound complications after coronary artery bypass graft surgery: bioengineered Apligraf, *Annals of Thoracic Surgery*, 78:673–8.

Alotto, D., Ariotti, S., Graziano, S., Verrua, R., Stella, M., Magliani, G. and Castagnoli, C. (2002) The role of quality control in a skin bank: tissue viability determination, *Cell and Tissue Banking*, 3:3–10.

Anderson, W. (2000) The Possession of Kuru: medical science and biocolonial exchange, *Comparative Study of Society and History,* 42:713–44.

Andrews, L. and Nelkin, D. (2001) *Body Bazaar: The Market for Human Tissue in the Biotechnology Age*, New York: Crown.

Andrews, P. and Benvenisty, N. (2005) Starting from building blocks: tissue engineering using stem cells, *Current Opinion in Biotechnology*, 16:485–6.

Andrews, P., Benvenistry, N., Mckay, R., Pera, M., Rossant, J., Semb, H. and Stacey, G. (2005) The International Stem Cell Initiative: Toward benchmarks for human embryonic stem cell research, *Nature Biotechnology*, 237:795–7.

Ang, G. (2005) History of skin transplantation, *Clinics in Dermatology*, 23:320–4.

Annandale, E. and Hunt, K. (2000) *Gender inequalities in health*, Buckingham: Open University Press.

Appadurai, A. (1986) Introduction: Commodities and the politics of value, in Appadurai, A. (ed.) *The Social Life of Things: Commodities in Cultural Perspective*, Cambridge: Cambridge University Press.

Ariss, R. (2003) Theorizing waste in abortion and fetal ovarian tissue use, *Canadian Journal of Women and Law*, 15:255–81.

Batty, D. (2007) Hybrid embryos get the go-ahead, *The Guardian*, Thursday 17 May.

Beeson, D. and Lippman, A. (2006) Egg harvesting for stem cell research: medical risks and ethical problems, *Reproductive Biomedicine Online*, 13, 4:573–9.

Bentley, G. and Minas, T. (2000) Treating joint damage in young people, *British Medical Journal*, 320,10:1585–8.

Bharadwaj, A. and Glasner, P. (2009) *Local Cells, Global Science: The Rise of Embryonic Stem Cell Research in India*, Abingdon: Routledge.

Bjorklund, A., Dunnett, S., Brundin, P., Stoessl, A.J., Freed, C.R., Breeze, R.E., Levivier, M., Peschanski, M., Studer, L. and Barker, R. (2003) Neural transplantation for the treatment of Parkinson's disease, *Neurology*, 2, 1:437–45.

Blackburn-Starza, A. (2010) Last update: Stem cell doctor exploited 'vulnerable' patients hearing finds, 19 April. See www.bionews.org.uk/page_54539.asp (last accessed 21 January, 2011).

Bock, K., Ibarreta, D. and Rodriguez-Cerezo, E. (2003) *Human Tissue-Engineered Products: Today's markets and future prospects*, EUR 21000 EN. Spain: European Commission Joint Research Centre, Institute for Prospective Technological Studies.

Boer, G.J. (1994) Ethical guidelines for the use of human embryonic or fetal tissue for experimental and clinical neurotransplantation and research, *Journal of Neurology*, 242, 1:1–13.

Bordo, S. (1995) *Unbearable Weight: Feminism, Western Culture and the Body*, Berkeley, California: University of California Press.

Bouchie, A. (2002) Tissue engineering firms go under, *Nature biotechnology*, 20:1178–9.

Bowker, G. and Starr, S. (2000) *Sorting Things Out Classification and its Consequence*, Cambridge, USA: Massachusetts Intitute of Technology.

Boyd, M., Flasza, M., Johnson, M., St Clair Roberts, J. and Kemp, P. (2007) Integration and persistence of an investigational human living skin equivalent in human surgical wounds, *Regenerative Medicine*, 2, 4:363–70.

Brittberg, M., Lindahl, A., Nilsson, A., Ohlsson, C., Isaksson, O. and Peterson, L. (1994) Treatment of deep cartilage defects in the knee with autologous chondrocyte transplantation, *New England Journal Medicine*, 331, 14:889–95.

Brown, N. (2003) Hope against hype – accountability in biopasts, presents and futures, *Science Studies*, 2:3–21.

Brown, N. and Kraft, A. (2006) Blood ties: banking the stem cell promise, *Technology Analysis and Strategic Management*, 18, 3/4:313–27.

Brown, N. and Webster, A. (2004) *New Medical Technologies and Society: Reordering Life*, Cambridge: Polity Press.

Brown, N., Kraft, A. and Martin, P. (2006b) The promissory past of blood stem cells, *BioSocieties*, 1:329–48.

Brown, N., Rappert, B. and Webster, A. (2000) *Contested Futures: A Sociology of Prospective Techno-science*, Aldershot: Ashgate.

Brown, N., Faulkner, A., Kent, J. and Michael M. (2006a) Regulating hybrids: 'making a mess' and 'cleaning up' in tissue engineering and transpecies transplantations, *Social Theory and Health*, 4:1–24.

Brusselaers, N., Monstrey, S., Vogelaers, D., Hoste, E. and Blot, S. (2010) Severe burn injury in Europe: a systematic review of the incidence, etiology, morbidity, and mortality, *Critical Care*, 14, 5:R188.

Bullock, A., Higham, M. and MacNeil, S. (2006) Use of human fibroblasts in the development of a xenobiotic free culture and delivery system for human keratinocytes, *Tissue Engineering*, 12:245–55.

Burd, A. and Chiu, T. (2005) Allogeneic skin in the treatment of burns, *Clinics in Dermatology*, 23:376–87.

Busby, H. (2006) Biobanks, bioethics and concepts of donated blood in the UK, *Sociology of Health & Illness*, 28, 6:850–65.

Busby, H. (2010a) The meanings of consent to the donation of cord blood stem cells: perspectives from an interview-based study of a public cord blood bank in England, *Clinical Ethics*, 5:22–7.

Busby, H. (2010b) Trust, nostalgia and narrative accounts of blood banking in England in the twenty first century, *Health: An Interdisciplinary Journal for the Social Study of Health, Illness and Medicine*, 14, 3:369–82.

Busby, H. and Martin, P. (2006a) Pilot project on the multiple ethics of umbilical cord blood banking, unpublished report.

Busby, H. and Martin, P. (2006b) Biobanks, national identity and imagined communities: the case of UK Biobank, *Science as Culture*, 15, 3:237–51.

Busby, H., Hervey, T. and Mohr, A. (2008) Ethical EU law?: The influence of the European Group on Ethics in Science and New Technologies, *European Law Review*, 33:803–42.

Carsin, H., Ainaud, P., le Bever, H., Rives, J.E.A., Lakhel, A., Stephanazzi, J., Lambert, F. and Perrot, J. (2000) Cultured epithelial autografts in extensive burn coverage of severely traumatized patients: a five year single-center experience with 30 patients, *Burns*, 26:379–87.

Casper, M. (1994) At the margins of humanity: fetal positions in science and medicine, *Science, Technology and Human Values*, 19:307–23.

Casper, M. (1998) *The Making of an Unborn Patient. A Social Anatomy of Fetal Surgery*, London: Routledge.

Check, E. (2003) Parkinson's transplant therapy faces setback. http://cmbi.bjmu. edu.cn/news/0308/174.htm.

Cooper, M. (2003) Rediscovering the immortal hydra: stem cells and the question of epigenesis, *Configurations*, 11, 1:1–26.

Cooper, M. (2005) Regenerative medicine: stem cells and the science of monstrosity, *Medical Humanities*, 30:12–22.

Cooper, M. (2006a) Resuscitations: stem cells and the crisis of old age, *Body and Society*, 12, 1:1–23.

Cooper, M. (2006b) *Bodily Transformations – Tissue Engineering and the Topological Body*, Working Paper No. 12. University of East Anglia: Global Biopolitics Research Group. June.

Cooper, M. (2008) *Life as Surplus*, London: University of Washington Press.

Council Of Europe (1994) *Recommendation No. R(94) 1 of the Committee of Ministers to Member States on Human Tissue Banks*. Recommendation R(94)1.

Council Of Europe (2002) *Guide to Safety and Quality Assurance for Organs, Tissues and Cells*, Strasbourg: Council of Europe Publishing.

Cox, M.T.J. (2000) Biomaterials regulation: innovative products – outgrowing the European framework? *Action for Biomaterials*, May, 4.

Daniels, S. (2006) Stem Cell Sciences plc, *Regenerative Medicine*, 1, 5:721–6.

Davis, K. (1995) *Reshaping the Female Body*, London: Routledge.

Department of Health (2000) *Guidance on the Microbiological Safety of Human Organs, Tissues and Cells used in Transplantation*, Advisory Committee on the Microbiological Safety of Blood and Tissues for Transplantation, London: Department of Health.

Department of Health (2001) *A Code of Practice for Tissue Banks Providing Tissues of Human Origins for Therapeutic Purposes*, London: Medicine Controls Agency, Department of Health.

Department of Health (2002) *Stem Cell Research: Medical Progress with Responsibility. A Report from the Chief Medical Officer's expert group reviewing the potential of developments in stem cell research and cell nuclear replacement to benefit human health*, London.

Department of Health (2006) *Review of the Human Fertilization and Embryology Act: Proposals for revised legislation (including establishment of the Regulatory Authority for Tissue and Embryos)*, London.

Department of Health (2007a) *On the state of public health: Annual report of the Chief Medical Officer 2006*. 28175. Crown.

Department of Health (2007b) Human Tissue and Embryos (draft) Bill. CM7807.

194   *Bibliography*

Department of Health (2007c) *Abortion Statistics, England and Wales: 2006.* 2007/01.

Department of Health (2010) *Liberating the NHS: Report of the arms-length bodies review.* London: Department of Health.

Department of Trade and Industry (2000) *Excellence and Opportunity: a science and innovation policy for the 21st century,* London: The Stationery Office.

Derksen, M. (2008) *Engineering Flesh Towards Professional Responsibility for 'Lived Bodies' in Tissue Engineering,* Eindhoven: University of Twente, PhD thesis.

Dickenson, D. (2002) Commodification of human tissue: implications for feminist and development ethics, *Developing World Bioethics,* 2, 1:55–63.

Dickenson, D. (2005) Human tissue and global ethics, *Genomics, Society and Policy,* 1, 1:41–53.

Dickenson, D. (2007) *Property in the Body: Feminist Perspectives,* Cambridge University Press.

Dolynchuk, K., Hull, P., Guenther, L., Sibbald, R., Brassard, A., Cooling, M., Delorme, L., Gulliver, W., Bourassa, D., Ho, V., Kunimoto, B., Overholt, T., Papp, K. and Tousignant, J. (1999) The role of Apligraf in the treatment of venous leg ulcers, *Ostomy/Wound Management: The Journal for Extended Patient Care Management,* 45, 1:34–43.

Donato, R., Miljan, E., Hines, S., Aouabdi, S., Pollock, K., Patel, S., Edwards, F. and Sinden, J. (2007) Differential development of neuronal physiological responsiveness in two human neural stem cell lines, *BMC Neuroscience,* 8:36.

Dunnett, S. and Rosser, A. (2006) Stem cell transplantation for Huntington's disease, *Experimental Neurology,* 203: 279–92.

Dunnett, S. and Rosser, A. (2007) Cell transplantation for Huntington's disease Should we continue? *Brain Research Bulletin,* 72:132–47.

Eaglstein, W.H., Alvarez, O.M., Auletta, M., Leffel, D., Rogers, G.S., Zitelli, J.A., Norris, J.E., Thomas, I., Irondo, M., Fewkes, J., Hardin-Young, J., Duff, R.G. and Sabolinski, M.L. (1999) Acute excisional wounds treated with a tissue-engineered skin (Apligraf), *Dermatological Surgery,* 25, 3:195–201.

Ehrich, K. (2010) *'Ethical frameworks for embryo donation': Ethics and Policy Workshop on Restricted Consent.* Wellcome Trust Biomedical Ethics Programme Project Grant No. 081414. King's College London.

Ehrich, K., Williams, C., Farsides, B., Sandall, J. and Scott, R. (2007) Choosing embryos: ethical complexity and relational autonomy in staff accounts of PGD, *Sociology of Health and Illness,* 29, 7:1091–106.

Eriksson, L. and Webster, A. (2008) Standardising the unknown: practicable pluripotency as doable futures, *Science as Culture,* 17:1.

European Group on Ethics in Science and New Technologies (1997) *Opinion No. 10 on the Ethical Aspcts of the Fifth Research Framework Programme.*

European Group on Ethics in Science and New Technologies (1998a) *Opinion No. 12 on the Ethical Aspects of Research Involving the Use of Human Embryo in the Context of the Fifth Framework Programme.*

European Group on Ethics in Science and New Technologies (1998b) *Ethical Aspects of Human Tissue Banking – Opinion of the European Group on Ethics in Science and New Technologies to the European Commission.* Online publication no. 11, 21 July 1998, available at: http://ec.europa.eu/bepa/european-group-ethics/docs/avis11_en.pdf

European Group on Ethics in Science and New Technologies (2000) *Opinion of the European Group on Ethics in Science and New Technologies to the European*

Commission: Opinion No. 15 on the Ethical Aspects of Human Stem Cell Research and Use, available at: http://ec.europa.eu/bepa/european-group-ethics/docs/avis15_en.pdf

European Group on Ethics in Science and New Technologies (2002) *Ethical Aspects of Patenting Inventions Involving Human Stem Cells – Opinion of the European Group on Ethics in Science and New Technologies to the European Commission*. Online publication no. 16, available at: http://ec.europa.eu/bepa/european-group-ethics/docs/avis16_en.pdf

European Group on Ethics in Science and New Technologies (2007) *Opinion No. 22 on the Ethics Review of hESC FP7 Research Projects*.

European Medicines Agency (2010) Public statement on concerns over unregulated medicinal products containing stem cells, Press release, 16 April 2010, available at: www.ema.europa.eu/ (last accessed January 2012).

European Parliament (2004) *Directive 2004/23/EC: Setting Standards of Quality and Safety for the Donation, Procurement, Testing, Processing, Preservation, Storage and Distribution of Human Tissue and Cells*, London: The Office for Official Publications of the European Communities. 31 March.

Farrell, A. (2009) The politics of risk and EU governance of human material, *Maastricht Journal of European and Comparative Law*, 16, 1: 41–60.

Faulkner, A. (2009a) *Medical Technology into Healthcare and Society: A Sociology of Devices, Innovation and Governance*, Basingstoke: Palgrave.

Faulkner, A. (2009b) Regulatory policy as innovation: constructing rules of engagement for a technological zone of tissue engineering in European Union, *Research Policy*, 38:637–46.

Faulkner, A., Geesink, I., Kent, J. and FitzPatrick, D. (2003) Human tissue engineered products – drugs or devices? *British Medical Journal*, 326, 7400:1159–60.

Faulkner, A., Geesink, I., Kent, J. and FitzPatrick, D. (2008) Tissue-engineered technologies: scientific biomedicine, frames of risk and regulatory regime building in Europe, *Science as Culture*, 17, 2:195–222.

Faulkner, A., Kent, J., Geesink, I. and FitzPatrick, D. (2006) Purity and the dangers of regenerative medicine: regulatory innovation of human tissue-engineered technology, *Social Science & Medicine*, 63:2277–88.

Fink, S., Schumacher, J., Ellias, S., Palmer, E., Saint-Hilaire, M., Shannon, K., Penn, R., Starr, P., VanHorne, C., Kott, H., Dempsey, P., Fischman, A., Raineri, R., Manhart, C., Dinsmore, J. and Isacson, O. (2000) Porcine xenografts in Parkinson's Disease and Huntingdon's Disease patients: preliminary results, *Cell Transplantation*, 9:273–8.

Fletcher, R., Fox, M. and McCandless, J. (2008) Legal embodiment: analysing the body of healthcare law, *Medical Law Review*, 16, 3:321–45.

Foucault, M. (1973) *The Birth of the Clinic: An Archaeology of Medical Perception*, London: Tavistock.

Fox, M. (2009) The Human Fertilisation and Embryology Act 2008: tinkering at the margins, *Feminist Legal Studies*, 17,3:333–44.

Fox, R.S.J. (1992) *Spare Parts: Organ Replacement in American Society*, Oxford: Oxford University Press.

Franklin, S. (2006a) Embryonic Economies, *BioSocieties*, 1:71–90.

Franklin, S. (2006b) The Cyborg embryo: our path to transbiology, *Theory, Culture and Society*, 23, 7–8:167–87.

Franklin, S. (2007a) *Dolly Mixtures: The Remaking of Genealogy*, London: Duke University Press.

Franklin, S. (2007b) 'Crook' pipettes: embryonic emigrations from agriculture to reproductive biomedicine, *Studies in History and Philosophy of Biological and Biomedical Sciences*, 38:358–73.

Franklin, S. and Lock, M. (eds) (2003) *Remaking Life & Death: Toward an Anthropology of the Biosciences*, School of American Research.

Franklin, S. and Roberts, C. (2003) Innovative health technologies: the social life of the embryo, available at: www.lancs.ac.uk/fass/sociology/papers/roberts-franklin-social-life-of-embryo.pdf (last accessed 16 January 2012).

Franklin, S. and Roberts, C. (2006) *Born and Made: An Ethnography of Preimplantation Genetic Diagnosis*, Princeton University Press.

Franklin, S., Hunt, C., Cornwell, G., Peddie, V., Desousa, P., Livie, M., Stephenson E. and Braude P. (2008) hESSCO: development of good practice models for hES cell derivation, *Regenerative Medicine*, 3, 1:105–16.

Freed, C., Greene, P., Breeze, R., Tsai, W-Y., Dumouchel, W., Kao, R., Dillon, S., Winfield, H., Culver, S., Trojanowski, J., Eidelberg, D. and Fahn, S. (2001) Transplantation of embryonic dopamine neurons for severe Parkinson's Disease, *New England Journal of Medicine*, 344, 10:710–19.

Freedlander, E., Boyce, S., Ghosh, M., Ralston, D.R. and MacNeil, S. (1998) Skin banking in the UK: the need for proper organization, *Burns*, 24, 1:19–24.

Gaskell, G., Stares, S., Allansdottir, A., Allum, N., Castro, P., Esmer, Y., Fischler, C., Jackson, J., Kronberger, N., Hampel, J., Mejlgard, N., Quintanilha, A., Rammer, A., Revuelta, G., Stoneman, P., Torgersen, H, and Wagner, W. (2010) *Europeans and Biotechnology in 2010. Winds of Change?* Brussels: European Commission.

Gilman, S. (1999) *Making the Body Beautiful. A Cultural History of Aesthetic Surgery*, Princeton: Princeton University Press.

Gisquet, E. (2008) Cerebral implants and Parkinson's disease: a unique form of biographical disruption? *Social Science & Medicine*, 67:1847–51.

Glasner, P. (2005) Banking on immortality? Exploring the stem cell supply chain from embryo to therapeutic application, *Current Sociology*, 53, 2:355–66.

Glasner, P. (2009) Cellular division: social and political complexity in Indian stem cell research, *New Genetics and Society*, 28, 3:283–96.

Golfier, F., Barcena, A., Harrison, M. and Muench, M. (2000) Fetal bone marrow as a source of stem cells for *in utero* or postnatal transplantation, *British Journal of Haematology*, 109:173–81.

Goodenough, T., Williamson, E., Kent, J. and Ashcroft, R. (2004) Ethical protection in research: including children in the debate, in Smyth, M. and E. Williamson, E. (eds) *Ethics, power, knowledge and consent*, Bristol: Policy Press.

Goodwin, M. (2006) *Black Markets: The Supply and Demand of Body Parts*, New York: Cambridge University Press.

Gordon, P., Yu, Q., Qualls, C., Winfield, H., Dillon, S., Greene, P., Fahn, S., Breeze, R., Freed, C. and Pullman, S. (2004) Reaction time and movement time after embryonic cell transplantation in Parkinsons Disease, *Archives of Neurology*, 61:858–61.

Gottweis, H. and Kim, B. (2010) Explaining Hwang-Gate: South Korean identity politics between bionationalism and globalization, *Science, Technology and Human Values*, 35, 4:501–24.

Graham, I., Harrison, M. and Nelson, A. (2003) Prevalence of lower-limb ulceration: a systematic review of prevalence studies, *Advances in Skin and Wound Care*, 16:305–16.

Grear, A. (2004) The curate, a cleft palate and ideological closure in the Abortion Act 1967 – time to reconsider the relationship between doctors and the abortion decision, *Web Journal of Current Legal Issues*, available at: http://webjcli.ncl.ac.uk/2004/issue4/grear4.html (last accessed January 2012).

Grear, A. (2010) *Redirecting Human Rights: Facing the Challenge of Corporate Legal Humanity*, Basingstoke: Palgrave Macmillan.

Halewood, P. (1996) Law's bodies: disembodiment and the structure of liberal property rights, *IOWA Law Review*, 81:1331–93.

Hall, S. (2003) *Merchants of Immortality: Chasing the Dream of Human Life Extension*, New York: Houghton Mifflin.

Hambly, K., Bobic, V., Wondrasch, B., Van Assche, D. and Marolovits, S. (2006) Autologous chondrocyte implantation: postoperative care and rehabilitation, *The American Journal of Sports Medicine*, 34:1020–38.

Haraway, D. (1995) *Simians, Cyborgs and Women: The Reinvention of Nature,* 2nd edn, London: Free Association Books.

Harding, K.G., Morris, H.L. and Patel, G.K. (2002) Science, medicine and the future: healing chronic wounds, *British Medical Journal*, 324:160–3.

Hayden, C. (2007) Taking as giving: bioscience, exchange and the politics of benefit-sharing, *Social Studies of Science*, 37:729–58.

Hierner, R., Degreef, H., Vranckx, J. and Garmyn, M. (2005) Skin grafting and wound healing – the 'dermato-plastic team approach', *Clinics in Dermatology*, 23:343–52.

Hitchcock, E.R. (1994) Stereotactic Neural Transplantation, *Stereotactic and Functional Neurosurgery*, 62:120–33.

Hoeyer, K. (2007) Person, patent and property: a critique of the commodification hypothesis, *Biosocieties*, 2, 3:327–48.

Hoeyer, K. (2009) Tradable body parts? How bone and recycled prosthetic devices acquire a price without forming a 'market', *BioSocieties*, 4:239–56.

Hoeyer, K. (2010a) After novelty: the mundane practices of ensuring a safe and stable supply of bone, *Science as Culture*, 19, 2:123–50.

Hoeyer, K. (2010b) An anthropological analysis of European Union (EU) health governance as biopolitics: the case of the EU Tissues and Cells Directive, *Social Science & Medicine*, 70, 12:1867–73.

Hoeyer, K. (unpublished) The anthropology of human boundary objects: exploring new sites for the negotiation of humanness, PhD thesis, University of Copenhagen.

Hoeyer, K. Nexoe, S., Hartlev, M. and Koch, L. (2009) Embryonic entitlements: stem cell patenting and the co-production of commodities and personhood, *Body & Society*, 15, 1:1–24.

Hogle, L. (1995) Standardization across non-standard domains: the case of organ procurement, *Science, Technology & Human Values*, 20, 4:482–500.

Hogle, L. (1996) Transforming 'body parts' into therapeutic tools: a report from Germany, *Medical Anthropology Quarterly*, 10, 4:675–82.

Hogle, L. (1999) *Recovering the Nation's Body. Cultural Memory, Medicine, and the Politics of Redemption*, London: Rutgers University Press.

Hogle, L. (2005) Enhancement technologies and the body, *The Annual Review of Anthropology*, 34:695–716.

Hohlfeld, J., De Buys Roessingh, A., Hirt-Burri, N. and Chaubert, P. (2005) Tissue engineered fetal skin constructs for paediatric burns, *Lancet*, 366:840–2.

Hollander, A., Dickinson, S. and Sims, T. (2006) Maturation and integration of tissue engineered cartilage implanted in injured and osteoarthritic human knees, *Tissue Engineering*, 12:1787–98.

Hollander, A., Macchiarini, P. and Gordijn, B. (2009) The first stem cell based tissue engineered organ replacement: implications for regenerative medicine, *Regenerative Medicine*, 4, 2:147–8.

Hopkins, N., Zeedyk, S. and Raitt, F. (2005) Visualising abortion: emotion discourse and fetal imagery in a contemporary abortion debate, *Social Science and Medicine*, 61:393–403.

Hopwood, N. (2000) Producing development: the anatomy of human embryos and the norms of Wilhelm His, *Bulletin of the History of Medicine*, 74:29–79.

Hopwood, N. (2005) Visual standards and disciplinary change: normal plates, tables and stages in embryology, *History of Science*, 43, 3, 141:239–303.

Hopwood, N. (2007) Making embryos and embryologists: wax models and portrait sculpture, *Philosophy and History of Science Seminar Serie*s, 20 February 2007.

House of Commons Science and Technology Committee (2005) *5th Report: Human Reproductive Technologies and the Law*, London: The Stationery Office.

House of Commons Science and Technology Committee (2007) *5th Report: Inquiry into Government Proposals for the Regulation of Hybrid and Chimera Embryos*, London: The Stationery Office.

House of Lords and House of Commons (2007) *Joint Committee on the Human Tissue and Embryos (Draft) Bill. Volume 1 Report*, HL Paper 169–1, HC Paper 630–1, London: Stationery Office.

House of Lords Select Committee on Science and Technology (2000) *3rd Report: Science and Society*, London: The Stationary Office.

Huang, Q., Pegg, D. and Kearney, J. (2004) Banking of non-viable skin allografts using high concentrations of glycerol or proplyene glycol, *Cell and Tissue Banking*, 5:3–21.

Hughes-Wilson, W. and Mackay, D. (2007) European approval system for advanced therapies: good news for patients and innovators alike (editorial), *Regenerative Medicine*, 2, 1:5–6.

Human Fertilisation and Embryology Authority (1994) *Donated Ovarian Tissue in Embryo Research and Assisted Conception, Public Consultation Document*, London: Human Fertilisation and Embryology Authority.

Human Fertilisation and Embryology Authority (2006) *Donating eggs for research: safeguarding donors*, London: Human Fertilisation and Embryology Authority.

Human Fertilisation and Embryology Authority (2007a) *Hybrids and Chimeras: A Consultation on the Ethical and Social Implications of Creating Human/Animal Embryos in Research*, London: Human Fertilisation and Embryology Authority.

Human Fertilisation and Embryology Authority (2007b) *Hybrids and Chimeras: A Report on the Findings of the Consultation*, London: Human Fertilisation and Embryology Authority.

Human Fertilisation and Embryology Authority (2011) *Donating Sperm and Eggs: Have Your Say*, available at: www.hfea.gov.uk/5605.html (last accessed 26 January 2011).

Human Tissue Authority (2006) Code of Practice 5: Removal, Storage and Disposal of Human Organs and Tissue. London: Department of Health.

Human Tissue Authority (2009) *Code of Practice 9: Research*, London: Human Tissue Authority.

Human Tissue Authority (2011) Human Tissue Authority statement on the Academy of Medical Sciences review of the regulation and governance of medical research, 3 February, available at: www.hta.gov.uk (last accessed March 2011).

Husing, B., Buhrlen, D. and Gaisser, S. (2003) *Human Tissue Engineered Products – Today's Markets And Future Prospects. Final report for Work Package 1: Analysis of the Actual Market Situation – Mapping of Industry and Products*, Karlsruhe, Germany: Fraunhofer Institute for Systems and Innovation Research.

Irwin, A. (2001) Constructing the scientific citizen: science and democracy in the biosciences, *Public Understanding of Science*, 10:1–18.

Jasanoff, S. (2005) *Designs on Nature: Science and Democracy in Europe and the United States*, Oxford: Princeton University Press.

Kearney, J. (2001) Clinical evaluation of skin substitutes, *Burns*, 27, 5:545–51.

Kearney, J. (2005) Guidelines on processing and clinical use of skin allografts, *Clinics in Dermatology*, 23:357–64.

Kearney, J. (2006) Yorkshire regional tissue bank – circa 50 years of tissue banking, *Cell Tissue Banking*, 7:259–64.

Kelly, S. (forthcoming) Fetal cells and maternal bodies: biological objects, subject categories, and trouble at the boundary, *Science as Culture*.

Kemp, P. (2006a) Cell therapy – back on the up-curve, *Regenerative Medicine*, 1, 1:9–14.

Kemp, P. (2006b) History of regenerative medicine: looking backwards to move forwards, *Regenerative Medicine*, 1, 5:653–69.

Kent, J. (2000) *Social Perspectives on Pregnancy and Childbirth for Midwives, Nurses and the Caring Professions*, Buckingham: Open University Press.

Kent, J. (2003) Lay experts and the politics of breast implants, *Public understanding of science,* 12, 4:403–22.

Kent, J. (2007) Producing neural stem cells, creating value in the fetal tissue economy, paper presented at European Sociological Association Conference, University of Glasgow, September.

Kent, J. (2008a) The fetal tissue economy: from the abortion clinic to the stem cell laboratory, *Social Science & Medicine*, 67:1747–56.

Kent, J. (2008b) Mapping the fetal tissue economy: an invisible human project, British Sociological Association Medical Sociology Annual Conference, University of Sussex, September.

Kent, J. (2009) Enabling Stem Cell Science in the UK? A short history of a regulatory maze, *Beyond Pattison: Stem Cell Initiative Final Conference*, Wellcome Trust, London, May 2009.

Kent, J. and Faulkner, A. (2002) Regulating human implant technologies in Europe – understanding the new regulatory era in medical device regulation, *Health, Risk & Society*, 4, 2:189–209.

Kent, J. and ter Meulen, R. (2011) Public trust and public bodies: the regulation and use of human tissue for research in the United Kingdom, in Lenk, C., Sandor, J. and Gordijn, B. (eds) *Biobanks and Tissue Research. The Public, the Patient and Regulation*, Dordrecht: Springer.

Kent, J., Faulkner, A., Geesink, I. and FitzPatrick, D. (2006a) Culturing cells, reproducing and regulating the self, *Body & Society*, 12, 2:1–24.

Kent, J., Faulkner, A., Geesink, I. and FitzPatrick, D. (2006b) Towards governance of human tissue engineered technologies in Europe: framing the case for a new regulatory regime, *Technological Forecasting and Social Change*, 73:41–60.

Keywood, K. (2000) More than a woman? Embodiment and sexual difference in medical law, *Feminist Legal Studies*, 8:319–42.

Kitzinger, J. (2008) Questioning hype, rescuing hope? The Hwang stem cell scandal and the reassertion of hopeful horizons, *Science as Culture*, 17, 4:417–34.

Kraft, A. (2008) From bedside to bench? Communities of promise, translational research and the making of blood stem cells, *Science as Culture*, 17, 1: 29–41.

Kraft, A. (2009) *Processes, Dynamics and Problems in Translation: A Comparative Study of Stem Cell Innovation in the UK: Full Research Report*, ESRC End of Award Report RES-350–27–0005, Swindon: ESRC.

Kraft, A. (forthcoming) Converging potential(s) reconsidered: the stem cell and cancer, *BioSocieties*.

Krishnamoorthy, l., Harding, K. and Al, E. (2003) The clinical and histological effects of Dermagraft in the healing of chronic venous leg ulcers, *Phlebology*, 18, 1:12–22.

Lakshmipathy, U. and Verfaille, C. (2005) Stem cell plasticity, *Blood Reviews*, 19:29–38.

Landecker, H. (2005) Living Differently in Time: Plasticity, Temporality and Cellular Biotechnologies, available at: www.culturemachine.net/index.php/cm/article/view Article/26/33 (last accessed March 2011).

Landecker, H. (2007) *Culturing Life: How Cells Became Technologies*, Massachusetts: Harvard University Press.

Langer, R. and Vacanti, J. (1995) Artificial organs, *Scientific American*, September: 100–03.

Lau, D., Ogbogu, U., Taylor, B., Stafinski, T., Menon, D. and Caulfield, T. (2008) Stem cell clinics online: the direct-to-consumer portrayal of stem cell medicine, *Cell Stem Cell*, 3:591–4.

Laurie, G. (2004) Patenting stem cells of human origin, *European Intellectual Property Review*, 59–66.

Lee, E. (2003) Tensions in the regulation of abortion in Britain, *Journal of Law and Society*, 30, 4:532–53.

Lee, E. and Jackson, E. (2002) The pregnant body, in Evans, M. and Lee, E. (eds) *Real Bodies: A Sociological Introduction*, London: Palgrave.

Leiva, R. (2006) A brief history of human diploid cell strains, *The National Catholic Bioethics Quarterly*, Autumn:443–51.

Lindahl, A., Brittberg, M. and Peterson, L. (2003) Cartilage repair with chondrocytes: clinical and cellular aspects, *Novartis Foundation Symposium*, 249:175–86.

Lindvall, O. and Bjorklund, A. (2004) Cell therapy in Parkinson's disease, *The Journal of the American Society for Experimental NeuroTherapeutics*, 1, 4: 382–93.

Lock, M. (2000) On dying twice: culture, technology and the determination of death, in Lock, M., Young, A. and Cambriosio, A. (eds) *Living and Working with the New Medical Technologies, Intersections of Inquiry*, Cambridge: Cambridge University Press.

Lock, M. (2001) The alienation of body tissue and the biopolitics of immortalized cell lines, *Body & Society*, 7, 2:63–92.

Lock, M. (2002a) Human body parts as therapeutic tools: contradictory discourses and transformed subjectivities, *Qualitative Health Research*, 12, 10:1406–18.

Lock, M. (2002b) *Twice Dead Organ Transplants and the Reinvention of Death*, Berkeley: University of California.

Lock, M. (2004) Living cadavers and the calculation of death, *Body & Society*, 10, 2–3:135–52.

Lock, M. and Farquhar, J. (eds) (2007) *Beyond the Body Proper: Reading the Anthropology of Material Life*, London: Duke University Press.

Lock, M. and Nguyen, V. (2010) *An Anthropology of Biomedicine*, Chichester: Wiley-Blackwell.

Lundin, S. (2002) Creating identity with biotechnology: the xenotransplanted body as the norm, *Public Understanding of Science*, 11:333–45.

Lundin, S. and Idvall, M. (2003) Attitudes of Swedes to marginal donors and xenotransplantation, *Journal of Medical Ethics*, 29:186–92.

Lundin, S. and Widner, H. (2000) Attitudes to xenotransplantation: interviews with patients suffering from Parkinson's Disease focusing on the conception of risk, *Transplantation Proceedings*, 32:1175–6.

Lynch, M. (2002) Protocols, practices, and the reproduction of technique in molecular biology, *British Journal of Sociology*, 53, 2:203–20.

Lysaght, M., (2004) Tissue engineering: the end of the beginning, *Tissue Engineering*, 10, 1:309–20.

Lysaght, M. and Crager, J. (2009) Origins (editorial), *Tissue Engineering*, 15, 7: 1449–50.

McHale, J., Habiba, M., Dixon-Woods, M., Cavers, D., Heney, D. and Pritchard-Jones, K. (2007) Consent for childhood cancer tissue banking in the UK: the effect of the Human Tissue Act 2004, *The Lancet Oncology*, 8, 3:266–72.

McLean, S. and Williamson, L. (2007) The demise of UKXIRA and the regulation of solid organ xenotransplantation in the UK, *Journal of Medical Ethics*, 33: 373–5.

Maienschein, J. (2002) What's in a name: embryos, clones and stem cells, *The American Journal of Bioethics*, 2, 1:12–19.

Marston, W.A., Hanft, J., Norwood, P. and Pollak, R. (2003) The efficacy and safety of Dermagraft in improving the healing of chronic diabetic foot ulcers: results of a prospective randomized trial, *Diabetes Care*, 26, 6:1701–5.

Martin, A. (2010) Microchimerism in the Mother(land): blurring the borders of body and nation, *Body and Society*, 16, 3:23–50.

Martin, E. (1994) *Flexible Bodies: Tracking Immunity in American Culture from the Days of Polio to the Age of AIDS*, Boston: Beacon Press.

Martin, P., Hawksley, R. and Turner, A. (2009) *The Commercial Development of Cell Therapy – Lessons for the Future? Survey of the Cell Therapy Industry and the Main Products in Use and Development: Part 1 Summary of Findings*, Nottingham: University of Nottingham.

Martin, P., Coveney, C., Kraft, A., Brown, N. and Bath, P. (2006) Commercial development of stem cell research technology: lessons from the past, strategies for the future, *Regenerative Medicine*, 1, 6:801–7.

Mason, C. (2007a) Regenerative medicine 2.0, *Regenerative Medicine*, 2, 1:11–18.

Mason, C. (2007b) Regenerative medicine the industry comes of age, *Medical Device Technology*, 18, 2:25–8.

Mason, J.K. and Laurie, G.T. (2001) Consent or property? Dealing with the body and its parts in the shadow of Bristol and Alder Hey, *The Modern Law Review*, 64, 5:710–29.

Maynard-Moody, S. (1995) *The Dilemma of the Fetus*, New York: St Martin's Press.

Medical Devices Agency (2002) *A Code of Practice for the Production of Human-Derived Therapeutic Products*, London: Department of Health.

Medical Research Council (2005) *UK Stem Cell Bank Code of Practice for the use of Human Stem Cell Lines*, London: Medical Research Council.

Medical Research Council (2006) *Stem Cells: MRC Research for Lifelong Health*, London: Medical Research Council.

Medical Research Council (2008) *Stem Cells: MRC Research for Lifelong Health*, London: Medical Research Council.

Medical Research Council (2010) *Code of Practice for the Use of Human Stem Cell Lines*, London: Medical Research Council.

Medical Research Council Working Group (2001) *Human Tissue and Biological Samples for Use in Research: Operational and Ethical Guidelines*, London: Medical Research Council.

Mithofer, K., Petersen, l., Mandelbaum, B. and Minas, T. (2005) Articular cartilage repair in soccer players with autologous chondrocytetransplantation: functional outcome and return to competition, *The American Journal of Sports Medicine*, 33:1639–47.

Mol, A. (2005) *The Body Multiple: Ontology in Medical Practice*, 2nd edn, United States: Duke University Press.

Morgan, D. and Lee, R. (1991) *Human Fertilization and Embryology Act 1990: Abortion and Embryo Research, the New Law*, London: Blackstone.

Morgan, L. (2002) 'Properly disposed of': a history of embryo disposal and the changing claims on fetal remains, *Medical Anthropology*, 21, 3–4:247–74.

Morgan, L. (2004) A social biography of Carnegie embryo No. 836, *The Anatomical Record (Part B.: New Anatomy)*, 276B:3–7.

Morgan, L. (2006) Strange anatomy: Gertude Stein and the avant-garde embryo, *Hypatia*, 21, 15:34.

Morgan, L. (2009) *Icons of Life: A Cultural History of Human Embryos*, California: University of California Press.

Morgan, L. and Michaels, M. (eds) (1999) *Fetal Subjects, Feminist Positions*, Philadelphia: University of Pennsylvania Press.

Moustafa, M. Simpson, C., Glover, M., Dawson, R.A., Tesfaye, S., Creagh, F.M., Haddow, D., Short, R., Heller, S. and MacNeil, S. (2004) A new autologous keratinocyte dressing treatment for non-healing diabetic neuropathic foot ulcers, *Diabetic Medicine*, 21, 7:786–9.

Mulkay, M. (1997) *The Embryo Research Debate Science and Politics of Reproduction*, Cambridge: Cambridge University Press.

Murdoch, A. (2007) *Information Leaflet for Egg Sharing for Research*, Newcastle: Newcastle Fertility Centre.

Mychaliska, G., Meuench, M., Rice, H., Leavitt, A., Cruz, J. and Harrison, M. (1998) The biology and ethics of banking fetal liver hematopoietic stem cells for in utero transplantation, *Journal of Paediatric Surgery*, 33, 2:394–9.

Nahman, M. (2006) Materializing Israeliness: difference and mixture in transnational ova donation, *Science as Culture*, 15, 3:199–213.

Nahman, M. (2008) Nodes of desire: Romanian egg-sellers, 'dignity' and feminist alliances in transnational ova exchanges, *European Journal of Women's Studies*, 15, 2:65–82.

National Institute for Health and Clinical Excellence (2000) *Guidance on the Use of Autologous Cartilage Transplantation for Full Thickness Cartilage Defects in Knee Joints*, London: NICE.

National Institute for Health and Clinical Excellence (2005) *The Use of Autologous Chondrocyte Implantation for the Treatment of Cartilage Defects in Knee Joints, Review of Technology Appraisal 16, Technology Appraisal 89*, London: NICE.

Naughton, G. (2001) An industry imperiled by regulatory bottlenecks, *Nature Biotechnology*, 19:709–10.

Nelson, T., Behfar, A. and Terzic, A. (2008) Strategies for therapeutic repair: The 'R3' regenerative medicine paradigm, *Clinical Translational Science*, 1, 2:168–71.

Niklason, L. (2001) Prospects for organ and tissue replacement, *Journal of American Medical Association*, 285, 5: 573–6.

Oakley, A. (1986) *The Captured Womb: A History of the Medical Care of Pregnant Women*, Oxford: Basil Blackwell.

Olanow, C., Goetz, C., Kordower, J., Stoessl, J., Sossi, V., Brin, M., Shannon, K., Nauert, M., Perl, D., Goldbold, J. and Freeman, T. (2003) A double-blind controlled trial of bilateral fetal nigral transplantation in Parkinson's Disease, *Annals of Neurology*, 54:403–14.

Omar, A., Mavor, A., Jones, A. and Homer-Vanniasinkam, S. (2004) Treatment of venous leg ulcers with *Dermagraft, European Journal Vascular and Endovascular Surgery*, 27, 6:666–72.

Ong, A. (2005) Ecologies of expertise: assembling flows, managing citizenship, in Ong, A. and Collier, S. (eds) *Global Assemblages: Technology, Politics and Ethics and Anthropological Problems*, Oxford: Blackwell Publishing.

Ong, A. and Collier, S. (eds) (2005) *Global Assemblages: Technology, Politics and Ethics as Anthropological Problems*, Oxford: Blackwell Publishing.

Panconesi, E. and Andreassi, L. (2005) Reality and fantasies: the psychology of skin transplants, *Clinics in Dermatology*, 23:325–31.

Parry, S. (2006) (Re)constructing embryos in stem cell research: exploring the meaning of embryos for people involved in fertility treatments, *Social Science and Medicine*, 62, 10:2349–59.

Patra, K.P. and Sleeboom-Faulkner, M. (2009) Bionetworking: experimental stem cell therapy and patient recruitment in India, *Anthropology & Medicine*, 16, 2:147–63.

Pattison, J. (2005) *UK Stem Cell Initiative: Report and Recommendations, Report 271735*. Crown Copyright; see www.dh.gov.uk/ab/UKSCI/index.htm (last accessed 16 January 2012).

Pavesio, A., Abatangelo, G., Borrione, A., Brocchetta, D., Hollander, A., Kon, E., Torasso, F., Zanasi, S. and Marcacci, M. (2003) Hyaluronan-based scaffolds (Hyalograft C) in the treatment of knee cartilage defects: preliminary clinical findings, *Novartis Foundation Symposium*, 249:203–17.

Petchesky, R. (1987) Foetal images: the power of visual culture in the politics of reproduction, in Stanworth, M. (ed) *Reproductive Technologies: Gender, Motherhood and Medicine*, Oxford: Blackwell.

Peterson, L. (1996) Articular cartilage injuries treated with autologous chondrocyte transplantation in the human knee, *Acta Orthopaedica Belgica*, 62, Suppl 1:196–200.

Peterson, L., Brittberg, M., Kiviranta, I., Akerlund, E. and Lindahl, A. (2002) Autologous chondrocyte transplantation. Biomechanics and long-term durability, *American Journal Sports Medicine*, 30, 1: 2–12.

Peterson, L., Minas, T., Brittberg, M., Nilsson, A., Sjogren-Jansson, E. and Lindahl, A. (2000) Two- to 9-year outcome after autologous chondrocyte transplantation of the knee, *Clinical Orthopaedics and Related Research*, 374:212–34.

Pfeffer, N. (1993) *The Stork and the Syringe: A Political History of Reproductive Medicine (Feminist Perspectives)*, London: Blackwells.

Pfeffer, N. (2008) What British women say matters to them about donating an aborted fetus for stem cell research: a focus group study, *Social Science and Medicine*, 66:2544–54.

Pfeffer, N. (2009a) Histories of tissue banking, in Warwick, R., Fehily, D., Brubaker, S. and Eastlund, T. (eds) *Tissue and Cell Donation: An Essential Guide*, Oxford: Wiley-Blackwell.

Pfeffer, N. (2009b) How work reconfigures an 'unwanted' pregnancy into 'the right tool for the job' in stem cell research, *Sociology of Health & Illness*, 31, 1:98–111.

Pfeffer, N. and Kent, J. (2006) Consent to the use of aborted fetuses in stem cell research and therapies, *Clinical Ethics*, 1, 4:216–18.

Pfeffer, N. and Kent, J. (2007) Framing women, framing the fetus: how Britain regulates the collection and use of aborted fetuses in stem cell research and therapies, *BioSocieties*, 2, 4:429–47.

Pianigiani, E., Ierardi, F., di Simplicio, F. and Andreassi, A. (2005) Skin bank organisation, *Clinics in Dermatology*, 23:353–6.

Piccini, P., Pavese, N., Hagell, P., Reiner, J., Bjorklund, A., Oertel, W., Quinn, N., Brooks, D. and Lindvall, O. (2005) Factors affecting the clinical outcome after neural transplantation in Parkinson's disease, *Brain*, 128:2977–86.

Pickstone, J.V. (2000) *Ways of Knowing: A New History of Science, Technology and Medicine*, Manchester: University of Chicago Press.

Pioletti, D., Laurent-Applegate, L., Zambelli, P. and Ito, K. (2007) Treatment of non-union fractures with fetal cell therapy, *European Cells and Materials*, 13:31.

Plomer, A. (2004) *Stem Cell Patents: European Patent Law and Ethics Report,* FP6 'Life Sciences, Genomics and Biotechnology for Health', SSA LSSB-CT-2004-005251, University of Nottingham.

Polkinghorne, J. (1989) *Review of the Guidance on the Research Use of Fetuses and Fetal Material*, Cm 762, London: HMSO.

Pollock, K., Stroemer, P., Patel, S., Stevanato, L., Hope, A., Miljan, E., Dong, Z., Hodges, H., Price, J. and Sinden, J. (2006) *A Conditionally Immortal Clonal Stem Cell Line From Human Cortical Neuroepithelium for the Treatment of Ischemic Stroke*, available at: http://dx.doi.org/10.1016/j.expneurol.2005.12.011, Oxford: Elsevier.

Price, D. (2005) The Human Tissue Act 2004, *The Modern Law Review*, 68, 5: 798–821.

Rabinow, P. (1999) *French DNA Trouble in Purgatory*, London: University of Chicago Press.

Rajan, K. (2006) *Biocapital The Constitution of Postgenomic Life*, Duke University Press.

Redfern Report (2001) *The Royal Liverpool Children's Inquiry Report*, London: HMSO.

ReNeuron (2006) *Annual Report*, Surrey: ReNeuron Group plc.

ReNeuron (2007) Press release: ReNeuron welcomes ground-breaking research on gene-based reprogramming of human stem cells, 21 November, available at: www.reneuron.com (last accessed February 2008).

Research Councils UK (2007) Stem cells: hope *not* hype, Research Councils UK.

Richter, W. (2007) Cell-based cartilage repair: illusion or solution for osteoarthritis, *Current Opinion in Rheumatology*, 19:451–6.

Roberts, C. (2008) Fluid ecologies: changing hormonal systems of embodied difference, in Smelik, A. and Lykke, N. (eds) *Bits of Life: Feminism at the Inter-sections of Media, Bioscience and Technology*, London: University of Washington.

Roberts, C. and Throsby, K. (2008) Paid to share: IVF patients, eggs and stem cell research, *Social Science and Medicine*, 66, 1:159–69.

Rose, N. (2007) *The Politics of Life Itself: Biomedicine, Power and Subjectivity in the Twenty-First Century*, Woodstock, Oxfordshire, UK: Princeton University Press.

Rose, N. and Novas, C. (2004) Biological citizenship, in Ong, A. and Collier, S. (eds) *Global assemblages: Technology, Politics and Ethics and Anthropological Problems*, Oxford: Blackwell Publishing.

Rosser, A. and Dunnett, S. (2006) Cell transplantation for Huntington's disease, *The Lancet Neurology*, 5:284–5.

Rosser, A., Barker, R., Harrower, T., Watts, C., Farrington, M., Ho, A., Burnstein, R., Menon, D., Gillard, J., Pickard, J. and Dunnett, S. (2002) Unilateral transplantation of human primary fetal tissue in four patients with Huntington's disease: NEST-UK safety report, *Journal of Neurology, Neurosurgery and Psychiatry*, 73:0–8.

Rowley, E. and Martin, P. (2009) *Barriers to the Commericialisation and Utilisation of Regenerative Medicine in the UK*, Nottingham: University of Nottingham.

Royal College of Nursing (2002) *Sensitive Disposal of Fetal Remains*, London: Royal College of Nursing.

Royal College of Obstetricians and Gynaecologists (2004) *The Care of Women Requesting Induced Abortion*, Evidence-based Clinical Guideline Number 7, London: Royal College of Obstetricians and Gynaecologists.

Royal College of Obstetricians and Gynaecologists (2005a) *RCOG Ethics Committee Response to Department of Health Review of the Human Fertilization and Embryology Act*, November, available at: www.dh.gov.uk/prod_consum_dh/groups/dh_digitalassets/@dh/@en/documents/digitalasset/dh_4132765.pdf (last accessed March 2011).

Royal College of Obstetricians and Gynaecologists (2005b) *RCOG Scientific Advisory Committee Response to Department of Health Review of the Human Fertilization and Embryology Act*, November, available at: www.dh.gov.uk/prod_consum_dh/groups/dh_digitalassets/@dh/@en/documents/digitalasset/dh_4132653.pdf (last accessed March 2011).

Sabo, D. and Gordon, D. (1995) *Men's Health and Illness, Gender Power and the Body*, London: Sage.

Salter, B. (2005) Transnational governance and cultural politics: the case of human embryonic stem cells and the European Union's Sixth Framework Programme, available at: www.york.ac.uk/res/iht/projects/l218252005/SalterTransnational Governance.pdf (last accessed March 2011).

Salter, B. (2006a) *Bioethics, Patenting and the Governance of Human Embryonic Stem Cell Science: The European Case*. Working Paper edn, available at: www.york.ac.uk/res/sci/ (last accessed August 2007).

Salter, B. (2006b) China and the global stem cell bioeconomy: an emerging political strategy? *Regenerative Medicine*, 1, 5:671–83.

Salter, B. (2007a) Patenting, morality and human embryonic stem cell science: bioethics and cultural politics, *Regenerative Medicine*, 2, 1:301–11.

Salter, B. (2007b) Bioethics and the global moral economy: the cultural politics of human embryonic stem cell science, *Science Technology Human Values*, 32:554.

Salter, B. and Salter, C. (2007) Bioethics and the global moral economy: the cultural politics of human embryonic stem cell science, *Science Technology Human Values*, 32:554.

Salter, B., Cooper, M., Dickins, A. and Cardo, V. (2007) Stem cell science in India: emerging economies and the politics of globalization, *Regenerative Medicine*, 2, 1:75–89.

Scheper-Hughes, N. (2001) Commodity Fetishism in Organs Trafficking, *Body & Society*, 7, (2–3):31–62.

Scientific Committee on Medicinal Products and Medical Devices (2001) Opinion on the State of the Art Concerning Tissue Engineering, available at: http://ec.europa.eu/food/fs/sc/scmp/out37_en.pdf (last accessed 12 Januray 2012).

Secretary of State for Health (2007) Government response to the report from the Joint Committee on the Human Tissue and Embryos (Draft) Bill, available at: www.official-documents.gov.uk/document/cm72/7209/7209.pdf (last accessed 12 Januray 2012).

Shakespeare, P. (2005) The role of skin substitutes in the treatment of burn injuries, *Clinics in Dermatology*, 23:413–18.

Shamblott, M., Axelman, J., Wang, S., Bugg, E., Littlefield, J., Donovan, P., Blumenthal, P., Huggins, G. and Gearhart, J. (1998) Derivation of pluripotent stem cells from cultured human primordial germ cells, *Developmental Biology*, 95, 1: 3726–731.

Sharp, L. (2007) *Bodies, Commodities and Biotechnologies: Death, Mourning and Scientific Desire in the Realm of Human Organ Transfer*, New York: Columbia University Press.

Sheldon, S. (1997) *Beyond Control: Medical Power and Abortion Law*, London: Pluto.

Sheldon, T. (2006) Dutch clinic is ordered to stop giving stem cell therapy, *British Medical Journal*, 333:770.

Sheldon, T. (2007) The Netherlands bans private stem cell therapy, *British Medical Journal*, 334:12.

Shildrick, M. (1997) *Leaky Bodies and Boundaries: Feminism, Postmodernism and (Bio)Ethics*, London: Routledge.

Shildrick, M. (2008a) Corporeal cuts: surgery and the psycho-social, *Body & Society*, 14, 1:31–46.

Shildrick, M. (2008b) The critical turn in feminist bioethics: the case of heart transplantation, *The International Journal of Feminist Approaches to Bioethics*, 1, 1:28–47.

Shildrick, M., Mckeever, P., Abbey, S., Poole, J. and Ross, H. (2009) Troubling dimensions of heart transplantation, *Medical Humanities*, 35, 1: 35–8.

Shin, G. and Griffin, G. (2002) Infectious hazards of autologous skin transplantation using murine 3T3 fibroblasts as a feeder layer for human keratinocyte culture: a report for UKXIRA and Department of Health, unpublished.

Siani-Rose, M. (2006) Theregen Inc., *Regenerative Medicine*, 1, 6:841–5.

Skloot, R. (2010) *The Immortal Life of Henrietta Lacks*, London: Macmillan.

Sleeboom-Faulkner, M. and Patra, P. (2009) The bioethical vacuum: national policies on human embryonic stem cell research in India and China, *Journal of International Biotechnology Law*, 5, 6:221–34.

Smelik, A. and Lykke, N. (eds) (2008) *Bits of Life: Feminism at the Intersections of Media, Bioscience and Technology*, London: University of Washington.

Smyth, L. (2002) Choice, rights and reproductive freedom, *Women's Studies International Forum*, 25, 3:335–45.

Snyder, B.J. and Olanow, C. (2005) Stem cell treatment for Parkinson's disease: an update for 2005, *Current Opinion in Neurology*, 18, 4:376–85.

Snyder, R. (2005) Treatment of nonhealing ulcers with allografts, *Clinics in Dermatology*, 23:388–95.

Spallone, P. (1989) *Beyond Conception: The New Politics of Reproduction*, London: Macmillan.

Stephens, N., Atkinson, P. and Glasner, P. (2008) Regulation at the UK Stem Cell Bank: securing the past, validating the present, protecting the future, *Science as Culture*, 17, 1:43–56.

Stevens, M. (2005) *In vivo* engineering of organs: the bone bioreactor, *Proceedings of National Academy of Sciences*, 102, 32:11450–5.

Supp, D. and Boyce, S. (2005) Engineered skin substitutes: practices and potentials, *Clinics in Dermatology*, 23:403–12.

Takahashi, K., Tanabe, K., Ohnuki, M., Nanta, M., Ichisaka, T., Tomoda, K. and Yamanaka, S. (2007) Induction of pluripotent stem cells from adult human fibroblasts by defined factors, *Cell*, 131, 5: 861–72.

Tatarenko, A. (2006) European regulations and their impact on tissue banking, *Cell Tissue Banking*, 7:231–5.

Taylor, P. Barker, R., Blume, K., Cattaneo, E., Colman, A., Deng, H., Fox, I., Gerstle, C., Goldstein, L., High, K., Lyall, A., Parkman, R., Pitossi, F., Prentice, E., Rooke, H., Sipp, D., Srivastava, A., Stayn, S., Steinberg, G., Wagers, A. and Weissman, I. (2010) Patients beware: commercialized stem cell treatments on the web, *Cell Stem Cell*, 7:1–7.

The Academy of Medical Sciences (2010) *Reaping the Rewards: a Vision for UK Medical Science*, Academy of Medical Sciences.

The Hinxton Group (2010) Statement on policies and practices governing data and materials sharing and intellectual property in stem cell science, November, available at: www.hinxtongroup.org (last accessed 22 February 2011).

Trommelmans, L., Selling, J. and Direickx, K. (2007) A critical assessment of the Directive on tissue engineering of the European Union, *Tissue Engineering*, 13, 4:667–72.

Turner, B. (1994) *Regulating Bodies: Essays in Medical Sociology*, London: Routledge.

Voltarelli, J., Couri,C., Stracieri, A., Oliveira, M., Moraes, D., Pieroni, F., Coutinho, M., Malmegrim, K., Foss-Frietas, M., Simoes, B., Fozz, M., Squiers, E. and Burt, R. (2007) Autologous nonmyeloablative hematopoietic stem cell transplantation in newly diagnosed Type 1 Diabetes Mellitus, *The Journal of the American Medical Association*, 297, 14:1568–76.

Wacjman, J. (2004) *Technofeminism*, Cambridge: Polity Press.

Wainwright, S., Williams, C., Michael, M., Farsides, B. and Cribb, A. (2006a) Ethical boundary-work in the embryonic stem cell laboratory, *Sociology of Health & Illness*, 28, 6:732–48.

Wainwright, S., Williams, C., Michael, M., Farsides, B. and Cribb, A. (2006b) From bench to bedside? Biomedical scientists' expectations of stem cell science as a future therapy for diabetes, *Social Science & Medicine*, 63:2052–64.

Wainwright, S., Williams, C., Michael, M., Farsides, B. and Cribb, A. (2007) Remaking the body? Scientists' genetic discourses and practices as examples of changing expectations on embryonic stem cell therapy for diabetes, *New Genetics and Society*, 26, 3:251–68.

Waldby, C. (2002a) Stem cells, tissue cultures and the production of biovalue, *Health: An Interdisciplinary Journal for the Social Study of Health, Illness and Medicine*, 6, 3:305–23.

Waldby, C. (2002b) Biomedicine, tissue transfer and intercorporeality, *Feminist Theory*, 3, 3:239–54.

Waldby, C. (2006) Umbilical cord blood: from social gift to venture capital, *Biosocieties*, 1:55–70.

Waldby, C. (2008) Oocyte markets: women's reproductive work in embryonic stem cell research, *New Genetics & Society*, 27, 1:19–31.

Waldby, C. and Cooper, M. (2008) The biopolitics of reproduction: post-Fordist biotechnology and women's clinical labour, *Australian Feminist Studies*, 23, 55:57–73.

Waldby, C. and Cooper, M. (2010) From reproductive work to regenerative labour: the female body and the stem cell industries, *Feminist Theory*, 11, 3: 3–22.

Waldby, C. and Mitchell, R. (2006) *Tissue Economies: Blood, Organs and Cell Lines in Late Capitalism*, Durham: Duke University Press.

Waldby, C. and Salter, B. (2008) Global governance in human embryonic stem cell science: standardisation and bioethics in research and patenting, *Studies in Ethics, Law and Technology*, 2, 1: 23, available at: www.bepress.com/selt/vol2/iss1/art12

Webster, A. (2007) *Health, Technology and Society: A Sociological Critique*, Basingstoke: Palgrave Macmillan.

Webster, A. and Eriksson, L. (2008) Governance-by-standards in the field of stem cells: managing uncertainty in the world of 'basic innovation', *New Genetics & Society*, 27, 2:99–111.

Weiss, G. (1999) *Body Images: Embodiment as Intercorporeality*, London: Routledge.

Widdows, H. (2009) Border disputes across bodies: exploitation in trafficking for prostitution and egg sale for stem cell research, *The International Journal of Feminist Approaches to Bioethics*, 2, 1:5–24.

Widdows, H. (2010) The Janus-face of new reproductive technologies: escaping the polarised debate, *International Journal of Public Theology*, 4:76–99.

Wilkinson, S. (2005) Biomedical research and the commercial exploitation of human tissue, *Genomics, Society and Policy*, 1, 1:27–40.

Wilkinson, S. and Kitzinger, C. (eds) (1994) *Women and Health: Feminist Perspectives*, London: Taylor & Francis.

Williams, C. (2005) Framing the fetus in medical work: rituals and practices, *Social Science and Medicine*, 60:2085–95.

Williams, C., Alderson, P. and Farsides, B. (2001) Conflicting perceptions of the fetus: person, patient, 'nobody', commodity? *New Genetics and Society*, 20, 3:225–38.

Williams, D. (1997) Engineering a concept: the creation of tissue engineering, *Medical device technology*, 10:8–9.

Williams, D. (1999) *The Williams Dictionary of Biomaterials*, Liverpool, UK: Liverpool University Press.

Williams, D. (2002) Clarity and risk: the challenges of the new technologies, available at: www.emdt.co.uk/?features.REF=43 (last accessed July 2007).

Williams, S. (2003) *Medicine and the Body*, London: Sage.

Williamson, E., Goodenough, T., Kent, J. and Ashcroft, R. (2004) Children's paraticipation in genetic epidemiology: consent and control, in Tutton, R. and Corrigan, O. (eds) *Genetic Databases: Socio-ethical Issues in the Collection and Use of DNA*, 1st edn, London: Routledge.

Young, I. (1984) Pregnancy embodiment: subjectivity and alienation, *Journal of Medicine and Philosophy*, 9:45–62.

Yu, J., Vodyanik, M., Smuga-Otto, K., Antosiewicz-Bourget, J., Frane, J., Tian, S., Nie, J., Jonsdottir, G., Ruotti, V., Stewart, R., Slukvin, I. and Thomson, J. (2007) Induced pluripotent stem cell lines derived from human somatic cells, *Science*, 318:1917–20.

Zhu, N., Warner, R., Simpson, C. Glover, M., Hernon, C., Kelly, J., Fraser, S., Brotherston, T., Ralston, D. and MacNeil, S. (2005) Treatment of burns and wounds usign a new cell transfer dressing for the delivery of autologous keratinocytes, *European Journal of Plastic Surgery*, 28:319–30.

Index